# DevOps Patterns for Private Equity

Technology organization strategies for high performing software investments

**DAVE MANGOT**

# DevOps Patterns for Private Equity

Technology organization strategies for high performing software investments

DAVE MANGOT

# Reviews of *DevOps Patterns for Private Equity*

*"My association with Dave, be it as a collaborator on the 'Funcast' or as a guide addressing our portfolio companies, has always been an enriching experience. He manifests a practical, yet seasoned approach when it comes to advocating for DevOps in companies of any size. This book serves as a testament to his expertise, providing you with a comprehensive understanding of DevOps in a manner only Dave can deliver."*

**Jim Milbery, Operating Partner, Parker-Gale and co-host of the Private Equity Funcast podcast**

*"This book is your guide, skillfully blending technological details with strategic wisdom. It offers a practical approach to leveraging DevOps as a catalyst for value creation in an ever-changing environment."*

**Nii Ahene, Chief Strategy Officer, Tinuiti**

*"Dave is one of a small group of people I've met in my career who can talk about deep technical issues and business impact interchangeably, which is a must-have skill for anyone working in private equity. This book distills complex topics into bite-sized chunks that you can read and apply quickly."*

**Scott Barstow, Operating Partner**

*"No longer a simple toolset, 'DevOps Patterns for Private Equity' positions DevOps as the beating heart of a productive engineering culture. It decisively illustrates how targeted, meaningful efforts will always outshine scattered, task-oriented actions."*

**Kathy Keating, CTO, Advisor, & Executive Coach**

**DevOps Patterns for Private Equity**

Copyright © 2023 Mangoteque LLC

ISBN: 979-8-9888401-0-7 (Paperback)
ISBN: 979-8-9888401-1-4 (E-Book)

First printing edition 2023.

# Contents

---

## KEY

Business Outcomes

Previously Published

An Illustrative Story

# Preface

*Make the hard things easy, so you can work on harder things.*

I was leading the DevOps transformation at Salesforce (and writing this book has shown me how formative *that* was) at about the time that the first real DevOps book, *The Phoenix Project*, was released. One of the authors, Gene Kim, used to give talks at conferences to recruit people to the cause. I'd heard his speech a number of times, but one of the most special times to me was when, as a result of the work I'd done, he gave that talk to the entire Technology & Product organization at Salesforce.

At the end of his presentation, Gene would always talk about being on a mission to improve the lives of thousands of IT professionals around the world. His words have always stuck with me.

My career has spanned more than a dozen companies, and I've always tried to improve things for both the company and my co-workers. When the DevOps movement began in earnest, I was drawn to its emphasis on both technical excellence and humanism. When I decided to go into business for myself, I discovered that, instead of just being able to improve the lives of people at the company where I worked, I now had the opportunity to help folks enjoy work more, make more meaningful contributions to the business, and do some of the best work of their lives, all the while having less stress, more time with their families and friends, and greater satisfaction.

By emphasizing both technical excellence and empowering humans as the creative, smart, and wonderful people that they are, I'm able to improve the lives of thousands of IT professionals too! When engineers are happier at work, they produce better business outcomes like customer satisfaction. When they produce better business outcomes, leadership is able to deliver on their promises. When leadership is able to deliver on their promises, investors are happier. It's one of those unique situations where the outcomes are win–win–win.

Which leads to the purpose of this book — to help you create systems in portfolio companies that lead to win–win–win situations.

This is not a how-to technology book. This is a book that ties technology organization choices to business objectives. One of the things I have always appreciated, that is core to the DevOps movement, is the fact that it has always emphasized tying engineering outcomes to business outcomes. The "first way of DevOps" from *The Phoenix Project*, which emphasizes systems thinking, starts with the business and ends with the customer. There is a deep recognition that, while many of us enjoy "nerding out" on the engineering aspects, there are very real business drivers at stake.

One of my favorite DevOps stories is from 2008, when a network administrator for the city of San Francisco was actually *jailed* for refusing to provide the administrative passwords for his network gear.[1] He hadn't wanted anyone to ruin his beautiful, perfect creation. However, engineers are not paid by their organizations to make beautiful creations. He didn't understand that the technology organization exists to solve real customer's problems and to help their business succeed.

It's my hope that this book will help you to shine a light on the connection between engineering perspectives and business outcomes. By the time you finish, you should have an understanding of how to better recognize the different engineering situations you see, whether during due diligence, during a board meeting, after purchase or acquisition, or just when facing an especially unfamiliar situation. After recognition, you should understand how that particular situation fits into the overall goal of getting good at delivering software, which

---

1  https://abcnews.go.com/Technology/story?id=5432751&page=1

makes businesses "twice as likely to meet or exceed their organizational performance goals."[2]

The things that I've written about in this book are what I see when I work with companies, regardless of their level of maturity. The things we examine are from the perspective with which I look at them. They are the things that are holding the organization back from achieving its objectives. I hope that in the pages ahead you'll learn to see these patterns for yourself and be able to understand what's required to change the system.

# Why Patterns?

I've held a belief for many years that wisdom is just pattern recognition writ large. When we're younger, we've seen something maybe a few times. It's not as easy to recognize any one thing as something we've seen before. There may be subtle differences or the context may be different.

One thing humans are naturally good at is pattern recognition. When you've seen something only a few times, it might be hard to recognize. When you've seen that thing dozens, hundreds, or thousands of times, it becomes much easier to recognize. I believe that as we get older and collect more life experience, we don't *know* more than someone who may have learned something in school or in a book, but we get better at recognizing things that we've seen before, what the expected outcomes will be, and how much control we have over a situation.

In this book we'll discuss patterns that I've seen often in portfolio companies. Organizations saying they're "agile" because a few engineers went to Scrum Master training, that prioritize work by what's in front of their noses, that are afraid to change their infrastructure because they might lose customer data. But just as we talk about patterns and anti-patterns in engineering, there are patterns that can help us escape from the obstacles that hold us back. These DevOps patterns help us to break through these impediments and move up to the next level. These patterns are the solutions.

---

2   *Accelerate State of DevOps* 2019 pp. 5.

# Why Private Equity?

One of the things I like most about the private equity folks is that they're what I would call *serious*. They have an investment thesis, clear objectives, and time constraints in which to achieve their milestones. The types of firms that I enjoy working with are driven, analytical, not afraid to make change (even big change), not afraid of the cutting edge, and they make decisions in the service of their objectives. They invest heavily at the start of the holding period and improve the operation of the business in areas like finance, business development, and technology, which helps the company to grow over the next few years. Then they exit at a higher level than when acquired. Some portcos are reacquired after a number of years by the same firm because they're ready to move to a higher level once again.

An investor once told me that I get a call when a company's growth has outpaced its maturity. Applying DevOps principles allows these companies to match their maturity to their growth. I constantly encourage the engineering teams I work with to *make the hard things easy* so they can work on harder things. By doing so, they start moving from slow ticketing systems and halting delivery to work that enables growth and flow. They take the step up to the next level. Achieving operational excellence in the technology organization (just like in sales or finance) is a great match for DevOps, which teaches how to achieve high-performing teams.

Ask any of my close friends what motivates me to do something. It's not "because it's there," or because someone said I can't do it, but because it's hard. I love working on hard problems. When my kids were young they discovered that I go to work and solve puzzles all day long and they would ask me why. I think of it as one of life's greatest joys.

Organization problems, technology problems, implementation problems. Those are the things that get me out of bed in the morning. Have you bought a few companies and combined them together and found that the developers say the operations folks are lazy and dumb, and the operations folks complain that the developers don't understand what they're asking for? Perfect, that's my cup of tea. I love to work in organizations that are multiples of Dunbar's number (the theoretical number of personal connections any one human can realistically maintain) in size, where communication is strained, and things are constantly delivered late. Thankfully, because of the profile of companies that private equity companies tend to buy, and what they do with them, portfolio companies have these kinds of problems in spades!

## Like *Barbarians at the Gate*?

Are there "bad" private equity firms out there that treat people poorly and predominantly make investments that harm society and the planet? Sure. There are approximately 10,000 private equity firms in the United States alone. It's simply not possible that every single one will be ethical. Stories abound in the press about this private equity firm that did awful things to healthcare, or that one that preys on the real estate market. I hope those companies fail, and I'm particularly choosy about my partnerships.

I choose to partner with firms that want to level up their portfolio companies to get better returns, not just look for the laziest way to make money. That work sounds boring anyway. I like hard problems.

## Who Is This Book For?

This book is both a business book and a technology book. Therefore, it's best suited for people who work at the intersection of business and technology. They can be:

- Operating partners with backgrounds in software engineering

- Operating advisors like former CTOs

- Portfolio company CTOs

It's also helpful to have read *The Phoenix Project*.

Does this mean the book won't be useful for others? Of course not. The goal of the book is to demonstrate the connection between technology organization choices and business outcomes. However, the folks listed above are well incentivized to make that connection happen.

# Who Am I?

Like a lot of people I've worked with, I became fascinated at a young age with what computers could do. I was getting paid to write code by the time I got to college. I graduated with a degree in Cognitive Science, which is basically computer science and cognitive neuropsychology mixed together (I was studying machine learning decades ago). I try to bring elements of my psychology background to my work.[3]

I moved to the Bay Area during the first dotcom boom and had no idea such a thing was happening. When employers found out that I was pretty good with computers, jobs weren't hard to come by. Over the years I bounced around to different jobs until I led a globally distributed team running the content delivery network (CDN) for Cable & Wireless. Big scale and tons of hard problems, transforming the way a complex system was managed. I was hooked. I took some other jobs, but by 2009 I knew that production web operations was what I wanted to spend my time doing.

Along the way I'd picked up a lot of skills with security, coding, management, and infrastructure, and I became an architect in Infrastructure Engineering at Salesforce right about the time that the DevOps movement was taking off in earnest. I became heavily involved in that community while still designing and implementing a lot of the foundations of what runs Salesforce today.

In some respects, I feel lucky that I went to an enterprise software company like Salesforce instead of a Facebook or a Google. My clients are run like enterprise software companies! It's very familiar, just at a smaller scale.

---

3   https://youtu.be/M1mbl4kUskQ

In my last operational role I ran the global Site Reliability Engineering (SRE) organization for the SolarWinds cloud companies. This is where I got my first experience working with private equity.

I've never been afraid to break new ground, and I can move fluidly between different groups like development, operations, and executives. When I began consulting I realized that I didn't enjoy working with startups, which don't understand my message, or very large enterprises. My message resonates well with middle market companies that have hard problems and are interested in building a solid business through sustainable growth.

My job is not to solve my clients' problems. My job is to help my customers solve their own problems. I can't know as much about their business as they do, but I know how to get them to approach the problem in the right way. If you're interested in the approach, and not simple solutions, then this book will be good for you.

There are so many people to thank who have helped to get me to where I am today. If you and I have crossed paths during my career, you have almost certainly taught me something that I should thank you for. I'd particularly like to thank Nii Ahene, Bruno Connelly, my clients who have taught me so much, Joel Dolisy, Vadim Friedberg, Joe Kim, Gene Kim, John Willis and the early DevOps leaders, James Leppek, Andy Meaden, Jim Pennypacker, Steve Pereira, Barzel Segal, John Stauffer, Paul Zuber and the crew at Thoma Bravo, the Learning From Incidents (LFI) community, the Flow Collective, Chad Upton, my coach Rochelle Moulton, my technical sparring partner and all around amazing person Peter C. Norton, my incredible family, Peggy, Micah, Alex, and Lidya, who've heard me comment endlessly on this book, and anyone else I may have unwittingly forgotten.

I hope this book enables you to see things that can be improved in your own organization or investments. I hope it will give you the courage to make bold changes when necessary, knowing that only by making changes to the system itself can you achieve the outcomes you desire. I hope that you'll join travelers like Gene Kim and me as we improve the lives of thousands of IT professionals, and by extension millions of their family and friends, (and, OK, maybe some leaders and investors too!) around the world.

Dave Mangot
San Francisco, California
May 2023

# Introduction:
# Get Good at Delivering
# Software™

*"It is not big beating the small anymore, instead it is fast beating the slow."*[4]
— Gene Kim

4    https://twitter.com/Dynatrace/status/1359581280224956417

What if I told you that Blockbuster Video had a seven-year head start over Netflix on streaming video? What if I told you that before Prime Video, Prime Shipping, Google Docs, and Android phones, Amazon was a place to buy books and Google was just a search engine like Altavista? Altavista?

How were these companies able to enter new and existing spaces that had nothing to do with where they started and be successful? They were good at delivering software. Sataya Nadella says that "Every company is a software company." If you buy a tractor from John Deere, one of the most important things that gets updated on that tractor is not the oil filter, it's the software. There's a company with 3,500 engineers, 500 software releases a week, and 4,000 apps or services in production. That company is The Home Depot.

If "software is eating the world," as Netscape co-founder Marc Andreesen is famous for saying, and you want your company to do the eating instead of being eaten like Borders Books, Barnes & Noble, and Blockbuster video ... then you need to get good at delivering software.

# Situational Awareness

The economic historian Carlota Perez, building on some definitions from Joseph Shumpeter, explains that we are currently in the sixth modern technological revolution.[5] She argues that new paradigms replace old ones, which are marked by a "turning point" at which the new paradigm becomes fully dominant. The current revolution is the Age of Information and Computers, and the new paradigm of software is a good explanation for the rise of companies like Netflix, Google, and Amazon. Being good at software enables them to enter and dominate spaces that were once controlled by other players.

The American Enterprise Institute proposes that "At the current churn rate, about half of today's S&P 500 firms will be replaced over the next 10 years as 'we enter a period of heightened volatility for leading

5   https://www.forbes.com/sites/stevedenning/2017/11/25/from-a-casino-economy-to-a-new-golden-age-carlota-perez-at-drucker-forum-2017/?sh=4eb5728c3b4e

companies across a range of industries, with the next ten years shaping up to be the most potentially turbulent in modern history.'"[6] This turbulence sounds very much in line with the ideas of Dr. Perez.

Being good at delivering software in this new age will be table stakes for any company that wishes to survive. According to the Dell Digital Transformation report in a survey of technology leaders, "Almost 1 in 3 are worried their organization may not survive the next couple of years."[7] Does your company have the systems it needs to be able to survive and thrive on the other side of the turning point?

The Home Depot, a home improvement store, has made massive investments in technology for the past few years. According to *Forbes*, The Home Depot uses "new technology tools that let store workers see a fuller picture of a shopper's spending history and offer them insights for upselling. The company's e-commerce operation grew more than 20 percent last year." This means the pace of competition is actually accelerating! The same Dell report showed that the Digital Adopters segment jumped by 16 percent between 2018 and 2020, compared to only 9 percent for the previous two-year period.[8]

Finally, former Treasury Secretary Larry Summers has said, "the economy of the future is likely to be 'Schumpeterian,' with creative destruction the norm and innovation the main driver of wealth. Products based on ideas — music, software, pharmaceuticals — require an enormous investment to develop but very little to keep making."[9] This fits in with Dr. Perez's thesis that mass production is a relic of the last technological revolution. We don't require heavy equipment and large factories for mass production in the software age.

The question you need to ask yourself is: Is your business, are your investments, ready to compete in this accelerating age of competition, creative destruction, and innovation? Do they have the tools, the

6   https://www.aei.org/carpe-diem/only-52-us-companies-have-been-on-the-fortune-500-since-1955-thanks-to-the-creative-destruction-that-fuels-economic-prosperity/

7   https://www.dell.com/en-us/dt/perspectives/digital-transformation-index.htm#pdf-overlay=/en-us/collaterals/unauth/briefs-handouts/solutions/dt-index-2020-executive-summary.pdf

8   https://www.delltechnologies.com/en-ae/collaterals/unauth/briefs-handouts/solutions/dt-index-2020-full-findings-report.pdf

9   https://www.wired.com/2002/03/schumpeter/

expertise, and the culture necessary to make it through the turning point? Or will they be left behind?

It's helpful to remember what Lean guru W. Edwards Deming said: "It is not necessary to change. Survival is not mandatory."

# Implications

Gene Kim, the author of *The Phoenix Project*, explains that it's not about being a large, established player anymore. Speed is the strategic differentiator at this point in the paradigm shift. But why?

Eric Reis, in his book *The Lean Startup*, describes the process for product development and finding product market fit as running a series of experiments. The goal is to "learn more quickly what works, and discard what doesn't." He advocates for what he calls MVPs, Minimum Viable Products. These aren't things that are fast to develop but just barely work. They are carefully crafted learning experiments that allow you to determine if you have an idea worth pursuing or whether you should acknowledge a low-cost failure and move on to the next hypothesis. If a company takes six months to run an experiment, they'll be a lot further behind in the learn-fast-and-succeed cycle than one that can run an experiment, or multiple experiments, daily.

Think back to our Netflix vs. Blockbuster example. Netflix is good at delivering software. It releases thousands of software experiments every day. Blockbuster possibly had some large, slow consulting company with antiquated development practices releasing experiments every few weeks or months. Who would have the competitive advantage in determining the best way to deliver streaming movies?

Two keys to success are to maintain a strong customer focus in the design and product phases of development and to work on a very short release cadence. Jez Humble, co-author of *Continuous Delivery*, explains: "One of the ways to blur the lines between design and delivery and to shrink the overall value stream lead time is for teams to practice hypothesis-driven development & TDD [test-driven development], working in very short cycles to get code (including experiments) into prod (using an experiment framework)."

There's an even more compelling reason to get code into production as quickly as possible: inventory. Toyota Manufacturing worked with Dr. Deming on process improvement in the 1970s and is famous for inventing just-in-time manufacturing. What problem was Toyota trying to solve with this invention? Inventory. Toyota realized that carrying excess inventory was expensive whether on the input or output side of the manufacturing process. In either case, they had to pay to store this inventory, which was wasteful and expensive. In addition, inventory was basically capital they had spent that they were unable to turn into revenue. If they had parts in a warehouse, they couldn't make money on those parts. If they had cars sitting outside a factory, but people couldn't buy those cars, they couldn't make money on those either.

With software, it's the same. If we pay our developers to write software, but that software is sitting in a code repository somewhere, that's an investment we've made in our business that we can't make money on. If that's the case, when do we make money? When the software is deployed in production. That's when our customers can see value in what we've developed and increase their subscription or recommend it to colleagues. That's also where we can run experiments to determine what will drive revenue or achieve product market fit. If there's a long delay between when the code is written and when it's deployed, it's just idle inventory. Capital invested without return.

# Whither DevOps

So how do we solve this problem? How do we get good at delivering software? What does that mean?

I've been involved with the international DevOps movement for approximately the past decade. I've spoken at DevOps conferences, written a DevOps course, and been featured in some of the DevOps books. This is how I view DevOps:

DevOps is a *collaboration* between development, operations, and *other teams* with the recognition that we are tasked with *achieving common business goals.*

The outcome of being good at DevOps is being good at delivering software. We know that the outcome for the entire business is improved business performance. How?

The *2019 Accelerate State of DevOps* report and the companion book *Accelerate: The Science of Lean Software and DevOps* determined that the "highest performers are twice as likely to meet or exceed their organizational performance goals." Consider this proposition if you haven't already: becoming good at delivering software *doubles your chances* of meeting or exceeding goals for customer satisfaction, market share, revenue, etc.

What the researchers discovered was that there are four key indicators that predict company performance in the highest performers: deployment frequency, lead time for changes, change failure rate, and time to recover. It's easy to see how these indicators fit in well with the idea of experiments as advocated in *The Lean Startup* and at Netflix. The first two indicators deal with how quickly and how often we can get our experiments into production to test a theory or a new way of doing things. They're also a great measure of how quickly we can deploy security fixes. Which leads us to our second two indicators, which have to do with the quality of our releases and how quickly we can recover from failure. If our site is down or degraded, it doesn't matter how compelling our product is, because we cannot realize returns on our investment if no one can buy the product.

Being able to walk the tightrope between running lots of experiments and keeping the site stable causes tension in software delivery. One example of a company that's good at delivering software doing this well is Google. Google is famous in the industry for having "error budgets" which encourage their products to have a *minimum* amount of downtime every month. That means that they want their teams to push as hard as possible to get new code into production (and run those experiments) up until the point where the budget is exhausted, and then deployments are slowed or stopped. You must be very good at delivering software to walk that tightrope safely and effectively.

But DevOps isn't just for development and operations — it's for *other teams* as well.[10] Every team is part of the value stream in some way, including security, design, legal, and marketing. I once worked with a marketing team that would only start on collateral (e.g., tweets, blog posts) *after* a feature was released. We had inventory that we'd paid for and released that was available for weeks that no customers knew about. We revamped our process so that the collateral was developed in coordination with the feature releases so that the investment that was made as a business could be realized, with customers engaging with new features weeks earlier than before. That led to shorter feedback cycles, greater customer satisfaction, increased revenue, etc. It was possible because we had a culture of participation and information sharing that let those who were closest to the information decide the best way to execute on the company's goals. Those in management explained the goals and targets and left it to the smart, empowered humans to participate and deliver the results necessary to achieve those objectives. In the DevOps movement, we refer to this as Westrum generative culture after a model of organizational culture developed by sociologist Dr. Ron Westrum. Many investors intuitively know that this culture is a key indicator of high-growth teams.

# Looking Forward

It's clear that at this turning point in the age of software, companies that are able to master the dominant paradigm of the moment will be the ones that survive and thrive in the new "golden age" of Dr. Perez. Today, it's not enough to be big, or loaded with capital. It comes down to execution and speed. Companies need to be good at delivering software.

If we follow the DevOps model, this can be accomplished in a way that's good for business and employees, which was why I formed my consulting firm Mangoteque. Employees who work like cogs in a machine, without respect and consideration, will never deliver the results we can achieve for our companies when everyone works together to deliver value to our customers.

---

10　https://blog.mangoteque.com/blog/2015/04/08/devops-across-the-enterprise-moving-past-dev-and-ops/

Adrian Cockroft, the architect of the Netflix move from data center to cloud, famously tells a story about C-level executives. After giving a talk about the amazing results that Netflix was able to achieve in its engineering organization and for the business, he would be mobbed by C-suite executives who would declare, "That's amazing! What results! Where were you able to find those talented people?" To which Adrian would reply: "I hired them away from you, empowered them, and then got out of their way."

# A Brief History of the DevOps Movement

*"People want to hear a devops secret and get disappointed when you tell them to chop wood and carry water."*

— Andrew Clay Shafer, early DevOps leader

It was an early DevOpsDays (a series of local conferences held around the world) in 2011, held in Mountain View, California. About 250 people were packed into the common areas of a data center facility. There were various sponsors, and the Salesforce presenter was wearing a t-shirt bragging about its uptime numbers while it was experiencing a major outage that day. People had flown in from all over the country to participate and collaborate on this new thing strangely called "DevOps" that was being discussed in Silicon Valley and online. As was already customary at these conferences, there were plenary talks, lightning talks (5-minute talks with 20 auto-advancing slides), and open spaces for discussion of various topics.

One of the early leaders of the DevOps movement, John Willis, gave a lightning talk about some of the health challenges he was overcoming, and the entire audience cheered and was supportive. What kind of a tech conference was this?? There were people putting together a band for people to jam in after the day's events had concluded. This felt less like a technical conference and more like the gathering of a tribe. I was fascinated and excited. I had attended the conference a year earlier at the LinkedIn offices, where a panel that included Gene Kim (who later went on to collaborate on *The Phoenix Project*, the most well-known DevOps novel in existence) were debating whether Quality Assurance departments had outlived their usefulness. Clearly these folks were not afraid to question the traditional orthodoxy about what was necessary to develop software. These were boundary spanners, agents of change, and deep thinkers.

People took turns at the front of the room to propose topics for the open spaces. If you were interested in their topic, you went to a designated area of the facility for the discussion after all the proposals were completed and the room assignments were made. You didn't need to speak and you didn't have to stay. Any amount of participation was the right amount of participation.

Soon there was a commotion. John Willis, who together with Damon Edwards had developed the CALMS (culture, automation, Lean, measurement, sharing) model of DevOps, was furious! Almost every proposed open spaces topic was about tools. "Don't you get it?" John exclaimed. "DevOps is about culture, not tools!" and he threatened to take the CALMS model and storm out.

The great DevOps debate about whether tools create culture or culture creates tools was born. Just another in a long list of challenges the movement would face in the early years as it struggled to find its footing.

# The Journey

That was more than 10 years ago, and the movement is no longer struggling to understand its methods. We've learned a lot. Books like *Accelerate* have given us the science behind the results we get. Sure, marketing and HR have muddied the waters a bit with DevOps tools and DevOps engineers, but where did it start? How has it changed? And where is it going?

## Origins

The moment of birth was much debated for a time within the DevOps movement. Ultimately, even the originators didn't really care who got credit, but it's generally agreed that there were two significant events that launched this movement:

- In 2009, at the annual Velocity Conference, Flickr's John Allspaw and Paul Hammond gave a talk entitled *10+ Deploys a Day: Dev and Ops Cooperation at Flickr*.[11] It would be difficult to overstate the gravity of this assertion at the time. People weren't merely in awe, they were in shock. At that time in the industry, releasing software was something that people did, at best, every few weeks. It was generally a major affair that took hours, required the coordination of many people, and was almost always done off-hours to minimize

11  https://youtu.be/LdOe18KhtT4

the potential impact to customers should something go wrong. And things often went wrong. To be able to deploy 10 times in a day (which is now routine at thousands of companies) was heresy and downright dangerous! It was an earthquake that shook the entire software industry.

- Soon after, Andrew Clay Shafer and Patrick Debois attended an Agile software delivery conference, and they both wanted to attend a planned discussion about Agile infrastructure. In one of the historical quirks of the movement, they didn't meet up for the discussion, but they ultimately found each other. They talked about Agile software delivery as it applied to infrastructure problems, something we now call infrastructure as code (IaC). At the time, Andrew was a co-author, with his college roommate, Luke Kanies, of a popular configuration management project (used to manage infrastructure) called Puppet. Patrick went on to coin the name DevOps and gave structure to the local DevOpsDays conferences, which would eventually be held in dozens of countries around the world.

## Traction

I continued to participate in the early DevOpsDays events, which were conveniently piggybacked onto the annual Velocity Conference, in Santa Clara, California, about web performance and operations. Many of the topics at the conference (co-chaired by John Allspaw) soon became major influences on the DevOps movement. I met John as he introduced *my* Velocity talk and got to know more and more of the early influencers over those years. I even started a DevOps discussion group at the software company where I worked at the time.

The DevOps movement continued to pick up steam and grow. The boundary spanners at the forefront continued to explore and bring new ideas. I moved on to Salesforce and discovered the need to move Salesforce from a lot of the old ideas that were holding it back. Because the company had been known as a major success in the industry in Agile adoption, smart leaders like Marc Benioff and Parker Harris were curious about anything that could keep Salesforce at the forefront of the industry. During my time at Salesforce, Gene Kim launched the DevOps Enterprise Summit, and I gave a talk at the very first one with Reena Mathew, a Salesforce Quality Engineering Principal Architect. Gene's mission was to spread DevOps beyond the local discussion groups and early adopting software companies into the larger enterprise software companies and beyond, a tradition he continues to this day with the Summit in the United States and Europe, his IT Revolution Press publishing company, and the dozens of contributions he makes to the movement each and every day.

## I'm Not a DevOps ... Are You an Agile?

I signed up for a new MeetUp group a few months ago, and as part of the process I was supposed to answer a few questions. One was, "Are you a DevOps?" I was struck by this. I'm pretty sure I knew what the question was supposed to mean, but I also knew that the question was nonsensical. Maybe a proper rephrasing would have been, "Are you an engineer who works in accordance with DevOps principles?" Maybe that was just too long and loaded to ask as a question. I feel like we're starting to lose control of the word "DevOps." Is that just natural for our industry? Is it a good thing or a bad thing? Maybe this means that DevOps is gaining traction, or maybe it means we've lost sight of why we're doing this.

On any given day my inbox is filled with job opportunities: "Looking for DevOps Engineers," or, one of my personal favorites, "Senior DevOps Engineer," or "Lead DevOps Engineer." I feel like sometimes our industry comes up with a term that gets dragged through the mud for a few years until perhaps the proper usage is found.

So then what about the term "DevOps Engineers"? Has Patrick Debois, has our industry, already lost control of our own term? I've been going to DevOpsDays for years, I've been talking to people and giving talks about DevOps, I've been trying to help salesforce.com through a DevOps transformation. In none of those interactions have we ever talked about DevOps engineers. When I give my Introduction to DevOps talks, I usually tell the audience to substitute the word "collaboration" where they see the word "DevOps" and they will be much closer to an instant understanding of what we're trying to accomplish with DevOps, even if they won't understand the "why" quite as quickly.

If you've ever listened to Gene Kim talk about DevOps, he talks about making life better for thousands of IT professionals. DevOps is definitely about getting rid of the "throw it over the wall" mentality between development and operations, but it's for the purpose of getting the business to focus on what's most important, being able to rapidly deliver value to the customer. In order to do that, Dev and Ops have to change too. Maybe those changes are a natural evolution. I remember the BOFH (B*stard Operator From Hell). I hope that, as an industry, we've put that behind us, we've evolved. The idea of a BOFH is antithetical to the DevOps movement. The BOFH said no to everyone, so it's no wonder people didn't want to collaborate with him or her.

The BOFH was fine for the days when we used to think it was a good idea to outsource our IT departments to cut costs. If your development wing is in the Philippines, and your operations department is in India, and they're run by two completely different companies, it doesn't matter all that much how well your Devs and Ops communicate, because they don't! You're already starting behind and will never catch up to companies practicing DevOps. DevOps isn't about outsourcing; DevOps is about insourcing. The problem is, no matter how much the engineer in us tries to escape it and turn everything into an algorithm, it's still all about people and how they communicate. The DevOps movement is a recognition of that fact. *Crucial Conversations*, a book about communication skills, teaches that when you're going to have an important conversation, you need to establish mutual purpose. That is DevOps in a nutshell — development and operations establishing mutual purpose. In DevOps, that purpose is to deliver the most amount of value to the business through streamlined processes; it's about always seeking to increase flow. Once that purpose is recognized, you're much more likely to have a successful conversation, and you're much more likely to have a successful business.

Obviously, our industry is in transition. Everyone is looking for "DevOps Engineers." Are you part of a movement that says that we can deliver more business value when people see delivery through Lean principles? Where the emphasis is on flow through the system and short feedback loops rather than silos and politics? It's people over process, a core tenet of the Agile Manifesto. Do I believe in people over process? Of course I do — that's why I believe in Agile. I think the lessons that the Agile development process teaches us not only make us better developers, network admins, and sysadmins, I think they also make us better people. The sprint retrospective may be about making the team better through a process of self-improvement, but it's also about remembering to improve ourselves. Any strides that we make as individuals, through better tooling or better communication, make the team better and benefit the business.

The sales and marketing folks are definitely having success using the "DevOps" term. When I signed up for the O'Reilly Velocity Conference, I was asked, "Do you primarily work in Web Operations, Web Performance, or DevOps?" Last week on Twitter, @PuppetLabs asked "Lots of DevOps jobs out there — and more on the way. How do you become a DevOps engineer and get in on this trend?" These are from two organizations that definitely get DevOps. Maybe we'll be calling all the jobs that are interesting "DevOps jobs" until the whole industry is operating with that business model. I just have to imagine that Toyota never advertised for "Kaizen Engineers" to work on the production line.

So when confronted with the question, "Are you a DevOps?" I answer the only way I know how. Someone was asking not whether I was qualified to fill a role but whether I believed in a movement. Whether flow was of the utmost importance. Whether communication was more important than silos. "Of course not, are you an Agile?"

# The Future

The DevOps movement has brought in so much knowledge and experience from things like Lean, software architecture, Agile, psychology, neuroscience, and management thinkers. And it's still changing, evolving, and growing.

Books, podcasts, whitepapers, and conference talks have contributed to the growth of the movement over the years. These include what I feel was the second earthquake to rock the industry, the IT Revolution book *Accelerate* by Dr. Nicole Fosgren, Jez Humble, and Gene Kim, which offered proof that high DevOps performance leads to better *business* outcomes for the companies that are proficient.

In my opinion, there are four areas that are at the forefront of the growth of DevOps currently, which in a few years will be normal parts of a movement that is continually adapting.

## Automated Governance

Since the early days of infrastructure as code, people have been fascinated by the possibility that computers would be able to have a major impact on the audit and governance process. Gene interviewed me about this at Salesforce in 2013!

Typically, auditors would arrive on site and review the governance guidelines and then have you pull a representative sample of machines to demonstrate that things are in fact as you say they are. With the dawn of configuration management, it was computers making sure that computers were configured correctly. An engineer would define a policy and the computers themselves would enforce that policy. The computers also kept records and logs that could be cryptographically signed as immutable proof of the changes.

Now, instead of going through a laborious audit once a year, there was continuous auditing — mathematically signed attestations that the fleet of deployed hardware and software was within compliance tolerances.

Over the years, people have expanded this philosophy to the software supply chain, to artifact creation, to deployments. Each was signed and recorded, documenting who did what and when. I expect this will become more common within the DevOps movement, and the old once-a-year model of examining systems we had to undergo for SAS-70 and the like will seem strange and rather quaint.

## Platform Engineering

Years ago I tweeted something to the effect that every job felt like I was repeatedly building a bespoke PaaS (Platform as a Service), like Heroku. Since the early days of the DevOps movement, Damon Edwards, co-founder of RunDeck, has been delivering the message, "Operations provides a platform."

With the release of the *Team Topologies* book, also by IT Revolution, the idea of Platform Teams (in actuality platform engineering teams) was codified and has been largely adopted throughout the industry.

In the early days of the movement, there was a lot of confusion about whether developers would need to learn to write infrastructure code and understand everything about the infrastructure down to a deep level. It's now generally accepted that a platform team builds a control plane that can be manipulated by the software development teams, which abstracts away a lot of the complexity required to build and operate infrastructure. For example, if you request an instance on AWS, you only worry about the instance and not about all the security, networking, electricity, virtualization technologies, etc. that are required to bring you that instance. The irony is, of course, that you still need to contend with many of those things at an instance level.

Nevertheless, the platform team builds abstractions on top of the underlying provider, which are customized to your specific business needs, whether for compliance or security, so that your stream-aligned teams can deliver quickly and safely. They're enabling the organization to "shift left." This will continue and become more mature.

# Value Streams

*Team Topologies* defines stream-aligned teams as yet another type. Generally they are the same as what we've typically known as development teams, but with the recognition that they're aligned with value streams. These streams are all the steps that it takes to deliver value to the customer. In order to get the most value from stream-aligned teams, we can perform value stream mapping, looking for waste in the system and trying to optimize our processes to be able to deliver value as efficiently as possible.

Value streams come from Lean and, along with value stream mapping, are currently being used by advanced DevOps practitioners to look at the software delivery process holistically and identify areas that can be improved. At the current bleeding edge of these efforts are value stream management and flow engineering, where, much like automated governance, the stages of the process are continually tracked and management decisions about how to react to changing conditions, either in the market or in the organization, are informed and guided by the resulting information.

Ultimately value streams are a complement to Agile (as evidenced by their adoption by the Scaled Agile Framework (SAFe)), which operates at the level of the overall system, as opposed to operating at the level of a specific development team.

# Resilience Engineering

With their book *Continuous Delivery*, Jez Humble and Dave Farley taught the DevOps movement about the importance of automated testing. No software should be released to customers without being tested first. The folks at Netflix realized that this could be problematic. Testing software is a measure of the quality characteristics of the software *at the time of testing*. However, with complex distributed systems nothing is static, interactions are variable, and the nature of the interactions themselves are changing all the time.

Netflix began a program of continuous testing called the Chaos Monkey, which was born from a larger concept, chaos engineering. In a nutshell, continually testing various types of failures ensures that, no matter what parts of the overall system change, the software will continue to live up to its stated obligations.

Over time, Casey Rosenthal at Netflix discovered that there were many others in the industry who were interested in these same concepts. Netflix began a series of Chaos Community Days for discussion with other companies like Facebook and Uber. As a result of these events, the chaos engineering community began to expand and mature. At the third Community Day, John Allspaw approached Casey, as well as Nora Jones of Netflix, and proposed that they enroll in the Human Factors & System Safety master's degree program at Lund University in Sweden. The movement soon discovered that there is an entire academic discipline devoted to learning from failures in medicine, aerospace, and other disciplines. There was a recognized need in the software engineering discipline to apply those lessons that are highlighted so vividly by chaos engineering.

The Learning From Incidents (LFI) community is growing by the day within and adjacent to the DevOps movement. People are beginning to look beyond "shallow metrics" like Mean Time to Recover (MTTR) and toward a deeper and more meaningful understanding of what causes these failures in the first place, and how we can use what we learn to become better at responding to untoward events when they occur.

The chaos engineering discipline has also been expanding into the idea of Continuous Verification by embracing and extending the ideas from continuous delivery and chaos engineering. With continuous verification, we're not only testing software performance but also security, and not merely with tests in an antiseptic environment. We're continually verifying that our production installations and software deployments don't have the security holes and vulnerabilities that they claim to not have.

One can easily hear the echoes of Automated Governance, where we began our forward-looking discussion.

# Technology Leadership and Business Outcomes

In the past few years, the DevOps movement has come into clear focus for the more advanced operators of private equity firms, who will use every advantage available to boost the performance of their software portfolios.

By embracing a movement that emphasizes both *speed* and *quality* and recognizes that they're not mutually exclusive but mutually reinforcing, these operators are able to accelerate the results in their investment theses in order to deliver better results, faster.

DevOps transformation is no easy task. Many of the software companies these capable firms buy have been around for a long time. They were not created as modern cloud native startups that internalize the DevOps concepts easily, and their existing methods are entrenched. Many are still following antiquated practices for developing software — creaky, old, and slow processes and procedures that feel like a weight on the organization — as they try to reconcile the demands of the investors with the new crop of engineers entering the workforce, who expect DevOps practices at their places of work.

I tell my clients, often, that this is hard work. I sometimes think they believe I'm just saying that to soften them up for some of the difficult decisions that must be made. In reality, it's often a struggle to learn to stop focusing on the tactical and begin with the strategic. After repeatedly being faced with the challenges of anticipating work, lining up initiatives, negotiating within and without the department for months on end, it begins to take hold.

These are humans we're trying to change and, like I often say, biology is messy. Nathaniel Barnes, CTO of MeridianLink, explains that, while your clients are undergoing these changes, you need to repeat the message often enough that even *you* are sick of hearing it. But while difficult, these DevOps practices are proven to work, even in organizations where other similar initiatives have failed.

These DevOps practices will be the focus of the remainder of the book. We'll talk about the struggles companies face, why they face those struggles, how to move toward a more DevOps practice, and how those factors affect *business outcomes*.

When DevOps principles are adopted, the engineers and the organization are happier because things work more smoothly. There's no more waiting for weeks or months for resources, or delivering resources hurriedly and with poor quality at the last minute. By emphasizing both speed and quality, better results are easier and it becomes a joy to come to work.

# The System of DevOps Patterns for Private Equity

**DevOps Patterns for Private Equity**

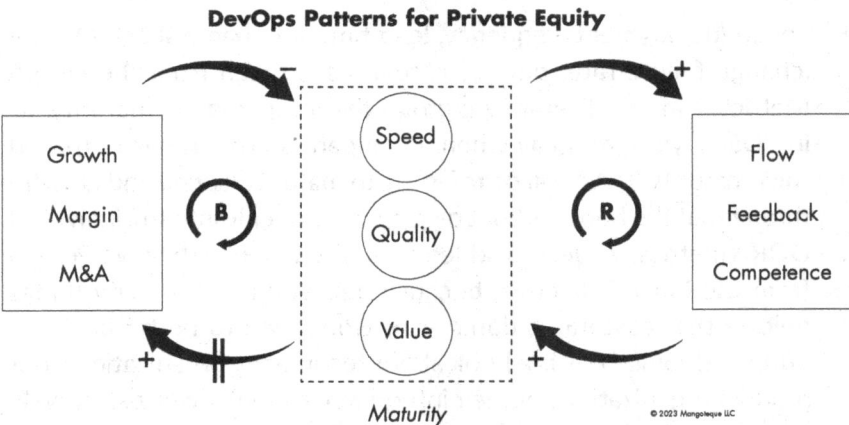

*Maturity*

© 2023 Mangoleque LLC

There is an expression in Spanish: *mañana en la mañana* (tomorrow in the morning), and this is a systems-level representation of systems-level representations. The diagram borrows from a few different frameworks and is intended to represent how the interplay between different parts of the system contributes to the overall effects.

The representation is in the style of a causal loop diagram and demonstrates how the different components of the system have effects on other components, paralleling that the decisions we make about our technology organizations have effects on business outcomes:

- The components on the right of the diagram (flow, feedback, competence) are taken from Gene Kim's blog post "The Three Ways: The Principles Underpinning DevOps" outlined by IT Revolution.[12] The first way emphasizes the performance of the entire system (systems thinking), always seeking to increase flow. The second way is concerned with shortening and amplifying feedback loops. The third way recognizes that repetition and practice lead to mastery. We will see these in patterns throughout the book as well as specific examples in the chapter about DevSecOps.

- The components on the left of the diagram are taken from *The Private Equity Playbook* by Adam Coffey. They represent three common levers that private equity uses in order to help its portfolio companies reach the next level so they can be ready for an exit.

- Speed (deployment frequency, lead time for changes) and Quality (change failure rate, time to recover) are taken from the DORA Metrics.[13] In the diagram we group them together as maturity (as in, "Dave, you get a call when a company's growth has outpaced their maturity."). We will be referring to maturity (speed and growth) throughout the book without digging more deeply into the individual DORA metrics. Value was added after discussion with Steve Periera from the Flow Collective, because delivering high-quality things quickly that customers don't want won't lead to better business outcomes. Since this is a book about technology organizations, not product organizations or their interplay, we won't dwell extensively on value — it's represented for the sake of completeness.

The right side of the diagram is a reinforcing (R) loop. For example, as flow increases or decreases, maturity increases or decreases. As speed increases, flow increases. Both these components reinforce each other like compounding interest, and the wheel spins faster and

---

12  https://itrevolution.com/articles/the-three-ways-principles-underpinning-devops/
13  https://cloud.google.com/blog/products/devops-sre/the-2019-accelerate-state-of-devops-elite-performance-productivity-and-scaling

faster. Similarly, the left side of the diagram is a balancing (B) loop. As growth outpaces maturity, speed and quality decrease. As speed and quality increase (or decrease), the ability to achieve the desired business outcomes follows suit, with a delay represented by the double bar.

I remember the first time I read *Thinking in Systems* by Donella Meadows. I was completely shocked to discover that other people saw the world the way I do. This systems-level thinking has been reinforced and revisited by management thinkers for over a hundred years.[14] We will reinforce and revisit this diagram throughout the book as we examine these systems.

Let's dig into how to deliver with speed, quality, and value together.

---

14  https://www.goodreads.com/book/show/1423027.The_Capitalist_Philosophers

# The Flow of Value

*What do you want to spend your engineering tokens on?*

"We need to hire more people," he said. Almost every time a team is overloaded and burned out, the first lever a manager wants to pull is to hire more people. Not because they're trying to gain more power and influence, but because there is "too much work." Throwing people at a problem can help, but even in inexpensive geographies, it's an expensive way to scale.

"What makes you say that?" I asked. He went on to explain. Like many private equity portfolio companies, this was actually a combination of multiple companies. The company that got to keep its name called it an acquisition. The other company called it a merger. Even though it had been years, most of the systems and work were still managed as before, separately.

As the company grew, it was able to hire more developers, but the hiring on the operations team hadn't come close to keeping pace with what was necessary, according to the vice president. With the addition of more developers came more work.

"How does work enter the system?" I asked. He explained that, for a long time, people would make requests in chat, in email, and also in the ticketing system. In the past few months, they required all work to come in through the ticketing system, though people still skirted that rule occasionally. Different people were responsible for different types of requests. For example, all networking tickets went to William. All Windows tickets went to Mary. All customer installs went to Bijal. Each responsible person worked as quickly as they could.

Over time, they realized there were different classes of work. Some required more of an expert, others were pretty mundane. They came up with an idea to create a smaller team that only handled the mundane tasks. This worked OK for a while but, as Aristotle said, "Nature abhors a vacuum," and soon everyone was fully extended all over again. The finance department didn't want to allocate any more budget for more people.

He wanted me to tell the CTO that they needed more people because, as a result of everyone running at 100 percent all the time, the staff were burned out and it was causing a lot of employee attrition. Fifty percent of the department had turned over in the last year alone. If they could just get a few more people, the team would be able to get over the hump and start burning down the backlog.

*System working as designed,* I wrote in my notebook, and began sketching an outline for a proposal.

One of the most common things I'm asked to help with in portcos is operations teams overloaded with, and trying to force all work through, ticketing systems. They try and work harder and smarter in a Sisyphean effort to handle all the work. The first thing I advise them is to give up. They'll never get through all the work. There will always be more things that they can work on than they can actually do. It's important that they always work on the things that are most important to the business (discussed in "Value Streams" below). This ruthless prioritization is what keeps teams focused on constantly delivering value. In essence, we have a limited number of engineering tokens, and we need to be very careful and deliberate about what we want to spend those engineering tokens on.

In this chapter, we'll talk about managing work in order to maximize our token spend.

# Toil

In *Site Reliability Engineering: How Google Runs Production Systems,* Vivek Rau describes toil as "the kind of work tied to running a production service that tends to be manual, repetitive, automatable, tactical, devoid of enduring value, and that scales linearly as a service grows."

Many times when companies are struggling to keep up with the work, there is a tremendous amount of toil. Even the vaunted Google engineers have toil, and they're supposed to keep it to less than 40 percent of the workload. However, in these companies, toil tends to dominate the work and engineers have trouble gaining traction.

These companies aren't trying to keep toil at less than 40 percent of the workload, they're trying to get 100 percent of the toil completed.

One can see in the definition that merely grinding through the work isn't the only solution. Rau says the work is "automatable," and these companies have humans doing a lot of work that a computer could do. I don't believe in doing work a computer can do. I believe in telling the computer what to do. If I'm reading a document on how to do a deployment by executing a series of steps, then the computer is telling me what to do. That's a situation that needs to change.

Humans get tired, they have bad days, they get into fights with their partners, they can't fall back to sleep at three in the morning. Computers will do the same things, over and over again, for days, weeks, and months on end and never complain and never miss a beat. It makes no sense to hire more humans to do work that a computer can do. The fact that toil is manual and repetitive makes humans even more likely to make mistakes or miss the signal in all of that noise. When they (inevitably) make a mistake, we blame the resulting problem on "human error," despite the fact that this is the exact outcome the system is designed to create.

W. Edwards Deming wrote about it in his 14 points:

> *Eliminate slogans, exhortations, and targets for the workforce asking for zero defects and new levels of productivity. Such exhortations only create adversarial relationships, as the bulk of the causes of low quality and low productivity belong to the system and thus lie beyond the power of the workforce.*

I once worked for a CEO who at the start of every quarter would ask everyone to give just a "little extra effort" for the next few months. *Every quarter!* After enough quarters of the same request, it became comical. We know that productivity gains achieved by improving the system far exceed any gains that can be achieved by asking for a little extra effort. This is the basic tenant of *The High Velocity Edge* by Dr. Steven Spear. Continual improvement. Continually improving the system.

If the work employees are doing is mostly tactical and devoid of enduring value, then we need to make efforts to improve the system, to better deliver value. We need to avoid trying to push more work through the same system, often by hiring more people. Why does Toyota allow people to write books about their approach to leadership and management? Because by the time the books are published, and the principles are adopted by others, Toyota will be far ahead of anything implemented, because Toyota continually improves. It's a learning organization. A learning organization doesn't try to push more work through the same system. It learns about the work from doing the work and then makes adjustments to allow more work to *flow* through the system.

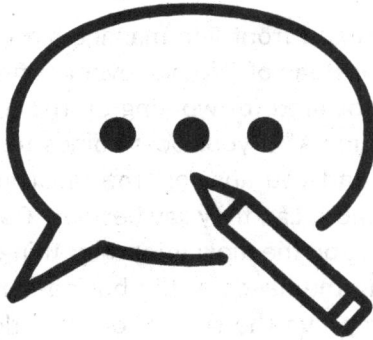

I worked with a client who went through a backlog of items and discussed when they should be scheduled. It was a very long list. Items could have a priority from 0 (critical) to 5 (lowest priority). At one point during the meeting I asked what the SLA was for an item classified as a 4. The response was: one year. If the team already had a full plate of work, and an item wouldn't be serviced for a year, at which point the business priorities could be different or the request wouldn't be relevant any more, why were we spending time discussing 4s and 5s? The meeting ended soon after.

# Value Streams

*"Maximize the work not done."*
— Old Agile adage

This adage doesn't mean to do as little as possible. It means to avoid working on things that don't bring as much value. As in Lean, the goal is to maximize value creation, therefore we need to work on the things that are most valuable to the business (and not devoid of enduring value, the way toil is).

I often tell engineers and front line managers a variation on the old saying about IBM — instead of "No one ever got fired for buying IBM," I say "No one ever got fired for working on the things that are most important to the business." If your boss comes to you and asks what you're working on, and you answer "the most important things to the business," what more can they say besides "Carry on." If a team is demonstrably working on the most important things, and their volume of work isn't meeting the needs of the business, having a discussion about whether to improve the system or add additional headcount makes sense.

Since leadership is generally (and rightly) remiss in continually adding headcount to avoid scaling "linearly as a service grows," they should first try to improve the system. But how? By looking at the system as a system. This is where value stream mapping can be useful.

| Design | → | Coding | → | Testing | → | Deployment |
|--------|---|--------|---|---------|---|------------|

Value stream mapping is a technique from Lean. We draw a map of all the stages from determining what the customer might want and that want being in the hands of the customer. This includes product specification, development, testing, and deployment. We make notations about how long things take at each stage of the process and look for ways to optimize them. There are a number of things in

Lean software development that are considered waste — they don't contribute to the value stream. These include:

- *Waiting*: Unit and security tests that run serially instead of in parallel that just cause delays

- *Handoffs*: The network engineers, then the system engineers, do their work, instead of infrastructure as code that does both

- *Context switching*: Too much Work in Progress (WIP) means engineers can't focus and complete work more quickly

- *Unnecessary complexity*: A change must be approved by many more people than are necessary to make it safe

Our example of discussing the priority 4 and 5 items above is a good one when looking at optimizing the value stream. There's an old joke in business about a company that doesn't allow employees to spend more than $100 without approval but has no problem calling 20 engineers into an hour-long meeting that accomplishes nothing. If the meeting was cut short when it was determined to deliver no value (waiting, unnecessary complexity), then waste would be eliminated from the system, and the engineers could use that time to deliver things that were most important to the business.

# Optimizing Work Flow

*"Hofstadter's Law: It always takes longer than you expect, even when you take into account Hofstadter's Law."*

– Douglas Hofstadter

What happens once we've spent time on optimizing the flow of work through the business? Are we done? How do we know when it's time to add more people to the team? To be able to make these judgements, we need to understand the delivery process of our teams and how well they're performing.

CTOs want less risk of something going off the rails and more predictability as to when something will be delivered.

With waterfall or PMO-driven companies, if a CTO asks a team when they can deliver project A, the team might say: two weeks. If they ask when they can deliver project B, the team will realize they need to pad the time for project B to account for project A. If the CTO asks when they can deliver project C, they will pad the time with A and B. Chance of delivering project C when they say it will be delivered? That's right, zero.

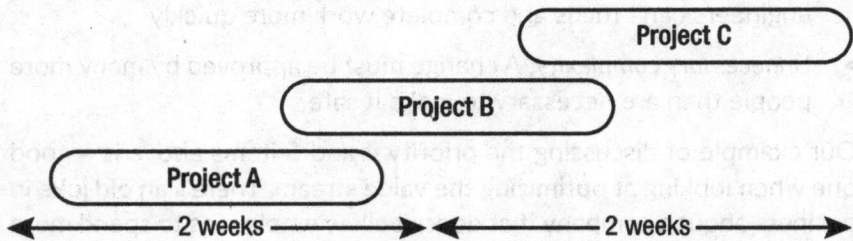

## Agile

There has to be a better way. There is: Agile prevents projects from going off the rails and gives more predictability as to when something will be delivered.

- *Why are things time-boxed in Scrum?* So that we can keep things from going off the rails by only committing to delivering the most valuable things within the sprint.

- *Why do we estimate story points and commit to delivering points in a sprint?* To give predictability as to when something will be delivered because we don't commit to more than the team can deliver within a time-box.

At some point the team will hopefully mature past the simplistics of a one-size-fits-all process (i.e., Scrum) and become truly Agile based on the things they have learned about the work itself, the work environment, and the team.

It's a lot easier to plan the release of that next big project or feature if you know when it will be delivered. Any leader will gladly accept a delivery date two weeks past their desired date if they can count on the fact that the work will be delivered on that date. The problems arise when they're continually promised dates that are never met, as in our waterfall example above. By continually delivering in small batches and planning with the most information we can have at the time of decision, we can reduce risk and improve predictability.

## What Agile Is Not

If Agile is a way to reduce risk and provide predictability, then what is it not?

- *Agile is not a tool to measure team productivity.* A story point doesn't really mean anything outside of a local context. A super junior team and a super senior team may both deliver 20 story points in a week, but you can be sure they don't deliver the same business value.

- *Agile is not a way to understand how much work is being done in an organization for planning purposes.* I've seen execs stand in front of a whiteboard for hours haggling over 20 points across an entire engineering organization based on what was done last quarter. When I asked what a single story point equated to, they answered: the amount of work an engineer can deliver in one day. I replied "junior or senior engineer?" There were a lot of blood-drained faces in the room that day. Additionally, if we worked in static systems, maybe this would make some sense, but people and contexts change constantly.

As stated, at some point the team should mature and become truly agile. I've seen portfolio companies where teams are allowed to mature only to the point where they can still provide Agile metrics to the executive dashboards! These teams are literally being held back from doing their best work so the ELT can feel they're in command and control.

To understand the flow of work through the system at an organization level, it's recommended to explore flow metrics (time, efficiency, etc.), which is an emerging field and outside the scope of this book.

## Scrum

Scrum is training wheels for Agile. Scrum is *not* Agile.

At many portfolio companies, executives ask why after sending a few people to Scrum Master training their teams are not agile. One of the main problems is that Scrum is not itself agile, Scrum is a method to *learn* Agile. Scrum as taught is extremely prescriptive, with many rituals to follow. If we remember that the goal of Agile is agility, it's impossible to imagine that one could be so severely constrained and still be "agile."

I'm not saying that the rituals aren't important and should be thrown away. The Adidas Runtastic team wrote a fantastic blog post called "How We Improved our Sprints When We Stopped Estimating Stories."[15] They describe how the team learned how to break up the work because of Scrum, and then were agile and adapted the work methodology to the team, instead of the team to the work methodology. That is *being agile* rather than *doing Agile* and is the desired and expected outcome from a team learning Scrum.

### Story Points

The most valuable thing teams learn from Scrum is how to break up the work. Being able to break up the work is useful for any kind of Agile methodology (not just Scrum) because of one of the primary goals of Agile: fast feedback.

If I'm building something off a specification from product management, I always want to know if I'm building the right thing. By the nature of being timeboxed in Scrum, I'll get the feedback that tells me I'm on the right track at the end of the sprint when demos are done. I can always demo what I've built earlier if desired. If the sprint is two weeks long, I'll have feedback in at *most* two weeks. Few things can be more heartbreaking for an engineer than to spend three months building something only to learn they've built the wrong thing and have it burned in front of them. This is damaging for the engineer and for morale, and it's downright costly for the business to have paid someone to build it.

15  https://www.runtastic.com/blog/en/no-more-estimates/

Story points require the work to be broken up to get this fast feedback. If the team has an average velocity of 20 story points per sprint, and they begin to tackle a 30-point story (or epic), they have a problem: 30 will not fit into 20. The story must be broken up. Of course, most teams have rules against stories larger than some number they decide, but you get the idea. Scrum forces teams to break work into smaller chunks. This way they can get fast feedback, can adjust more quickly and easily (i.e., be agile), and are encouraged to perform delivery in smaller batch sizes. After enough practice, the team will understand how to break up the work into reasonable chunks and can mature past the entire exercise of pointing their stories altogether. Scrum will have done its teaching job, and the team can spend that time on more valuable work.

## Retrospectives

In his book *The Unicorn Project*, Gene Kim extolls the virtues of the improvement of daily work. Improving work isn't something that just happens organically, especially if a team is focused on delivering features and fixing bugs. The retrospective makes time for the team to reflect and discuss. It is so rare that we set aside time in the workplace just to do *deep thinking*. What went well, what didn't go well, what can be improved? I have seen many teams put tremendous amounts of pressure on themselves at the end of every sprint to come up with something earth shattering as a result of the retrospectives. Instead they should be kind to themselves and focus on the improvement of daily work. James Clear, the author of *Atomic Habits*, explains that, by improving by 1 percent every day, we're 37 times better after one year![16]

Setting aside explicit time to be open and honest with one another also creates an opportunity to build psychological safety. In its landmark research study *The Five Keys to a Successful Google Team*,[17] Google discovered that its most successful teams were not the ones with the most senior engineers, or the most PhDs, but the ones with psychological safety. As they put it: "Team members feel safe to take risks and be vulnerable with each other." The other keys were

16  https://jamesclear.com/continuous-improvement
17  https://rework.withgoogle.com/blog/five-keys-to-a-successful-google-team/

dependability, structure and clarity, meaning of work, and impact of work. Setting aside time for retrospectives creates the spaces to create high-performing teams.

*Standups*

Another way that teams can build psychological safety and dependability is the daily standup. When running distributed teams, I would always leave the first five minutes of each standup for chit chat as people filtered into the video chat room. Why? So they can connect with each other as humans. To build psychological safety. What someone made for dinner last night? Sure. A big upcoming family event? Fair game. How their kid did at their soccer game yesterday? Absolutely. This chitchat builds understanding and connection. Understanding and connection build safety.

We still talked about what was done yesterday, what will be done today, and if anyone has any blockers they needed help clearing, but then, the meeting was over. Everyone understood what everyone else was working on and if they needed any help. No lingering around, no long discussions on a deep technical topic. It's called "standup" because you're supposed to *stand up* and, if you're not sitting, you're incentivized to be done with it and get on with your day.

# Kanban

Many of the best Scrum teams I've worked with eventually drift into Kanban. Kanban is a pull-based system that was popularized into a software work methodology by David J. Andersen, based on work from when he was at Microsoft.

Instead of time boxes, story points, and the like, work is prioritized and pulled from the top of a queue to make sure we're always working on the things most important to the business. Similar to how Scrum will only let us accept a certain number of points into a sprint, Kanban has a limit on how many things the team can work on simultaneously: WIP (pronounced *whip*), or work in progress.

Because complexity and context switching have costs, the team will be able to get more work through the system per unit time if they don't take on too much work. We can actually determine optimal WIP limits for a team.

In Kanban we measure two things: lead time and cycle time. Lead time is how long it takes from something being put on the queue until it's completed. Cycle time is how long it takes from when work is begun on an item until it is completed. By measuring these times, and experimenting with reducing (or raising) the WIP, we can determine how many simultaneous items a team can optimally handle.

One of my favorite things about this type of system is that it helps to make drastically clear the true capacity of the system. If I know that my team has a WIP limit of five, and my boss asks for a "special favor" that we take on extra work for the next few weeks, I know that will reduce the throughput of the team. I can ask in return, "Are you asking me to get less work done?"

# Making Work Visible

*"Every system is perfectly designed to get the results it gets."*
— W. Edwards Deming

We now understand that Agile is well suited to fast feedback and on-time delivery. We understand that examining the value stream can help us identify areas for improvement. But can they help one another? That is, are they complementary?

The answer is a resounding yes. Working in an agile manner allows us to visualize how work flows through the system, either on a Scrum board or Kanban wall. When Agile teams see work slow or get stuck, they *swarm* on the problem in order to unblock the work streams. Seeing these bottlenecks in the system can also have the benefit of highlighting the areas to examine in the value stream, looking for places to improve.

I was working with a client who would stage a number of software changes in a pre-production environment until the appointed release day. The tickets would stack up at the edge of the board. Not only did it look strange, but it also highlighted the fact that they were artificially creating big-bang releases and severely hampering their ability to get fast feedback from their customers. They were also increasing the risk for every release — as more changes stacked up, there was more chance for unintended interactions and failures.

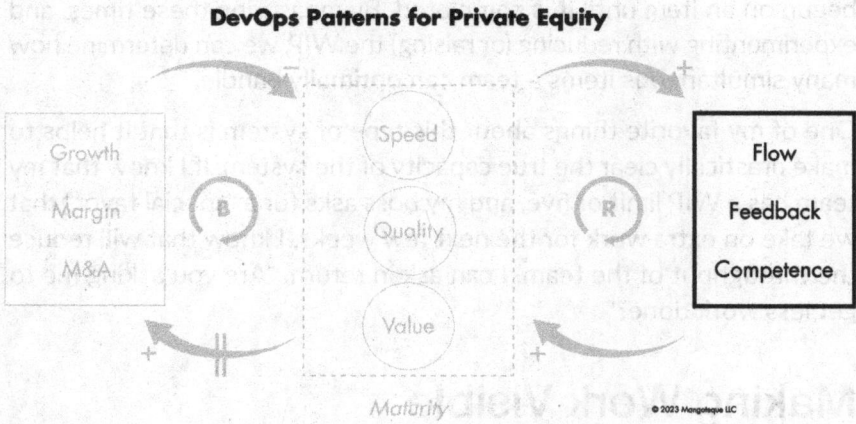

## Status Reports

Visualizing work flows has other advantages. We've all worked for a boss who would ask for a status report at the end of a release, knowing full well that they would give it a cursory view at best. When we make work visible, not only can we generate a report of what was done when, for any time period, but we can also see where work is in-flight, at any time, day or night. This eliminates the need for a project manager who continually chases after each team member asking for their status. This frees up the project manager to do much more valuable work coordinating large-scale initiatives across the company or investigating special projects where needed, instead of acting as a babysitter for adults.

# *Closing the Loop*[18]

In our story we have a sales guy, Bern. For the purposes of our story, Bern is good at his job. He is very responsive to his customers and tries very hard to give them what they want. He would also never throw something over the wall to his technical staff. Bern has been hearing from a lot of his customers that they would like feature X in his SaaS product. Bern talks to Product and finds out that many other customers besides his own are interested in feature X as well. So Bern and a member of the product team contact the head of engineering and ask her how long it would take for her teams to develop that feature.

Our head of Engineering, Tess, has a number of teams that practice Agile methodologies. They estimate that they can deliver feature X in two sprints at two weeks each, which means one month until it's in the hands of the customers. Everyone agrees, and Bern, being very responsive, explains to the customer that they should be able to purchase or use this new feature in about a month.

One of Tess's teams starts on the new feature at the beginning of their next iteration. Like all good Scrum teams, at the end of their first iteration, they hold the sprint retrospective where they demo what they have so far. During the course of the sprint, they have also been moving the stories that are part of the feature across the wall. During this time, Bern has had the ability to track the progress of the stories through the system

---

18  https://web.archive.org/web/20151004193259/http://blog.librato.com/posts/devops-across-the-enterprise

because of this transparency. On the day of the sprint retro, Tess's team records a short (< 5-minute) demo of each of the features they've been developing and posts the demos on the wiki page assigned to their team. Then they go home for the weekend.

Posting the demos for all to see accomplishes a number of things:

- Despite the fact that Bern is located on the other side of the planet from this Dev team, he can view the demo and comment on it and make suggestions.

- As a matter of fact, anyone in the organization — Sales, Product, or otherwise — can view the demo.

- By posting the demos, even those stories that aren't complete for the features, the Dev team is able to get fast feedback so they don't hand over a feature at the end that wasn't what the customer wanted.

- Bern can go to his customers and tell them that he's seen an early version of the feature and that it looks great (and that he perhaps made some suggestions for further improvement), so they understand the situation exactly.

When Tess's team is finally able to deliver the feature, it's exactly what the customer wanted. This is not just because her team, along with Ops, have spent a lot of time in the past few years implementing all kinds of optimizations to their system, it's because they made their work visible so that adjustments could be continually made.

# Anti-Patterns

There are a few things that I've seen at portfolio companies that, however well intentioned, dramatically reduce the flow of work through the system.

## Change Advisory Boards

A hushed murmur spread throughout the room when Dr. Nicole Fosgren, one of the authors of *Accelerate*, stood on stage and announced to the entire conference that "Change advisory boards are useless!"[19] Much like the Allspaw & Hammond talk years earlier, this was heterodoxy! In fact, she continued to explain, they are actually "negatively correlated with stability." The very thing they are supposed to maintain.

Change Advisory Boards (CAB) are part of the British government technology guidance Information Technology Infrastructure Library (ITIL). According to the guidance, when someone wishes to make a change in a production environment, they are required to prepare a document for presentation to the CAB describing the change, the risks involved, and the rollback mechanism. In theory, the board will review this document, and if they have multiple eyes to review the change, it will be safer.

To be sure, occasionally the board will find things to make a change seem safer. However, almost every major outage that happens is the result of something unexpected, not because someone was careless or didn't understand their document well enough. The person making the change is almost certainly the person with the most in-depth knowledge of the change they're making and the systems to which it is being applied. The expectation with a CAB is that either the person making the change will understand it better as a result of the process, or that people who don't understand it nearly as well will spot the unexpected occurrence and avert disaster.

Approval of the end result of our work by people with little understanding of the change itself is far too late in the process. Change advisory boards are trying to inspect quality into the product at the end, which is too late.

*"Inspection does not improve the quality, nor guarantee quality. Inspection is too late. The quality, good or bad, is already in the product. As Harold F. Dodge said, 'You can not inspect quality into a product.'"*

— W. Edwards Deming

---

19  https://youtu.be/DgpsX5yLXQw

There is an appropriate time for people to review the work. An engineering team will do code reviews. As we've seen, there are sprint demos. However, when I have had to present the end result of my work to a change advisory board, the board did not contribute anything but frustration.

I was going to deploy new file-sharing software to some equipment in Japan. I'd deployed this same software in the United States and Europe and was just trying to make sure that our network was covered geographically. It was required that I go before the board in order to justify the action I was taking and explain the steps I'd taken to make it safe. My work was just one of many changes that the board that I was explaining myself to had to approve in the hour allocated. The people on the board were all upper-level managers who didn't know the first thing about the software or the protocol it was supporting. Nonetheless, here I was, jumping through a worthless hoop that was purely an obstacle and contributed nothing to the business.

Contrast that with the work we did at Salesforce that's featured in *The DevOps Handbook*. There, each change that was proven to be verified by software to be safe was *automatically* classified as a standard change and would sail untouched through the change approval process. Only manual changes would have to be approved, and as Dr. Fosgren has described, even that had a negative effect.

Putting more knowledgeable engineers on the boards wouldn't be much of an improvement. As Deming describes, inspection is too late. We need to build quality in. The CAB lengthens the feedback loop, regardless of who's doing the reviewing at the end.

## Silos

Change advisory boards are an anti-pattern because of work not flowing smoothly through the system (waiting and unnecessary complexity in Lean), as are silos (handoffs).

When beginning an engagement with portfolio companies, I always ask to look at an organizational chart to see the reporting structures. If the Ops and Dev folks are reporting up through completely different structures, that's a red flag. Conway's Law (discussed below) says that this won't lead to the desired outcomes. Each silo will have its own objectives and won't think about the flow of work; they'll think about how to get "their part" done. I've seen so many portco engineers talk about "our problem" and "their problem." They're not actively trying to sabotage other teams' efforts, they just don't look at the system itself. As we liked to tell the security folks at Salesforce, "You'd be OK with the company going out of business as long as we were secure doing it!"

The customers don't care which part of the organization drops the ball, and finger pointing won't get the software delivered. I often explain to engineers and leaders that this is a software company, not a security company or a network security group building company, and as such, the goal is to deliver software.

When doing Assessments we always look for some kind of alignment exercise (e.g. V2MOM, OKR) that cascades down from leadership throughout the organization. Leaders set their goals and objectives, and each level of the hierarchy below them creates goals that support those objectives as well as their personal growth. This creates an organization that is aligned and focused on delivering the same objectives while looking at this delivery as a system.

When this alignment doesn't exist, it's common to see handoffs and people just trying to get work off their plate by prioritizing work based on whatever is in front of their face. This doesn't contribute to the value stream and results in halting delivery and missed deliverables.

Each team has an important part to contribute to the value stream if we can achieve both organizational and structural alignment.

# Conway's Law

*"Any organization that designs a system (defined broadly) will produce a design whose structure is a copy of the organization's communication structure."*
— Melvin E. Conway, computer programmer, 1967

I was talking with a SaaS company CEO who'd acquired a company in South Asia and was having a lot of trouble after the deal had closed. He explained that the Asian engineers operated separately from the American ones. The CTO spent a lot of time with the American engineers but relied on a manager in Asia to handle things with those engineers. He asked about things he could change in order to take better advantage of the new team and their existing product.

"What was the objective of the M&A activity?" I asked.

"What do you mean?"

"Well, you have two engineering teams and two products. Is it going to be two products fully integrated? Two separate products? One product? What are you trying to accomplish?" I replied.

"Oh, it's supposed to be one product with the functionality of both."

I explained that, with the current operating model, that objective was impossible because of Conway's Law. As long as the communication structure of the organization was such that the two products were kept separate, the system will continually resist integration. He needed to place the Asian and American programmers on teams together, and with both bringing their expertise, they'll be able to create a fully integrated product.

The longer the CTO refused to spend time with the new team, the longer this situation would persist, and the M&A activity would be for naught.

# Missed Deliverables

There are obviously many reasons why deliverables can be missed. However, when teams don't have an effective system they can work in, then the chances of missing the deliverables are even greater.

I've worked with many portfolio teams where the method of work prioritization is what I call "whatever is in front of my nose"; it's generally either FIFO (first-in, first-out) or LIFO (last-in, first out). This can be OK if a company is fine with treading water, but recency bias is a poor prioritization scheme. As more and more work comes in, teams simply try to push more and more work through the existing system, essentially treading water harder. But treading water harder will rarely lead to increased efficiency or faster time to delivery. It will also rarely lead to things being delivered on time. As we'll see when discussing Finance, those systems are optimized for 100 percent utilization. There's no slack in the system. The moment anything unexpected happens, the schedule is thrown off. The team is essentially fighting the inherent randomness in the system — what Deming called *common cause variation*.

If, on the other hand, we pad in too much time, we wind up promising estimates on top of estimates, which will only deliver when promised with massive amounts of luck — the stars aligning in just the right way that all our estimates were correct. What we typically see instead is that as the promised delivery date gets closer, engineering teams work more nights and weekends. Management can buy dinners and promise rewards in the future, but the damage is done. The teams lose faith in leadership, get tired of being repeatedly burned out, and find employment elsewhere.

# *Attrition*

We know the old expression: "Employees don't leave companies, they leave managers." When employees see their situation continually getting worse, or a bad situation staying the same, they leave. Regardless of market conditions, a portco with high attrition can be catastrophic.

Many of the senior engineers had been with the company for a long time, and despite a number of efforts to improve things, they were burning out. They would see long-time colleagues leave, and no one wishes to be the last one to abandon ship. Because these engineers had been around for a while and were experienced, when they ultimately left they were able to score a pretty nice payday. As the engineering saying goes: "There is no compression algorithm for experience." New engineers were simply not prepared to start where the departing engineers had left off. This can be a problem with offshoring as well.

As any production engineer knows, it takes time to come up to speed on a new system. How things are configured, what the conventions are for the environment, what the mores are for the culture. Even senior engineers aren't drop-in replacements, and humans are not simply fungible like fossil fuels. This can cause longer and more frequent outages, customers getting upset, and a ramp-up period of at least a few months while the new engineers come up to speed, and the morale across the organization takes a hit.

To prevent engineers from leaving, we need to give them interesting, meaningful, and impactful work that drives company growth in a collaborative environment. In short, succeeding on a team.

# Teams

*"If you want to go fast, go alone. If you want to go far, go together."*
— African proverb

"God bless America," she said.

"Excuse me?" I replied.

"God bless America. That's what we say when something happens here that makes us frustrated because we want things to change, but they don't change. We say God bless America."

With any large client engagement, I begin by spending a few weeks interviewing engineers about their work, the culture of the company, how they think things are going, and what they think could be improved. I usually schedule 30 minutes so it doesn't look too intimidating, but only the very shy engineers finish in that amount of time.

"Ok, that's good to know. What can you tell me about how things got to 'God bless America' levels of frustration?"

In the beginning things actually used to work pretty well. Then the company started growing, and growing. The team used to deal with that growth by working extra hours at night. One engineer told me that at this point they only got work done when no one was around, from 19:00–21:00. Then also on weekends. They felt like they kept trying to solve new things the old way.

Besides working more hours, I asked what they'd tried to change about the work itself. One engineer explained that he used to do "whatever was needed" but had started to focus on infrastructure monitoring because they'd barely gotten beyond simple "up/down" status for things. At least now they were collecting better metrics. He confided that they had no specific requirements for things and no real agreements to uphold, only things people had decided. Communication was suffering. "Basically, no one has time to do anything."

Each person was on an island by themselves. Developers wanted things tomorrow, Ops wanted to deliver in two weeks. They were definitely busy, but they weren't productive.

"There are too many environments, and it's all manual work," she said.

"Who's responsible for the environments?" I asked.

"No one is."

# Project vs. Product

*"The project-oriented management approach of 'bringing people to work' is not suited for complex knowledge work, like software delivery. High-performing software organizations have already learned that 'bringing work to the people' is more effective."*

— Dr. Mik Kersten, *Project to Product*

Project managers have been a part of many industries, including the software industry, for some time. As we saw in the previous chapter, their jobs have changed quite a bit in the past few years as Agile has become ascendant and the tracking of daily work and making it visible are now a standard part of the way we deliver software. There is an impedance mismatch in that project managers are generally assigned to manage *projects*, that is, work items of limited duration that have a well-defined start and end or acceptance criteria.

But the business of delivering software has no end. Software isn't "done" until all the running copies are deleted off all servers everywhere and it's not possible to restore any copies. Burying it all underground is a safer bet.

This is core to the problem of delivering software with a project mindset. Projects are crossed off the list upon what is considered completion, and then things that are needed to support that software (e.g., patches, monitoring, incident response) continue on indefinitely. Software (and infrastructure) has a lifecycle, hence Software Delivery Lifecycle (SDLC). Project work is ill suited to support a lifecycle. This is the case for the teams that write the software as well at those that support those teams.

Development teams will support the software they write as a product, i.e., a software product. This means that they are continually releasing patches, updates, or improvements, always trying to make the software better. They make a better product over time.

Operations teams fall behind Development when given tasks. Upgrade the memory, add monitoring to this queue, create another database replica. In essence, these tasks are projects. The Ops team continually tries to execute one project after another. This may have worked when the company was small, but as the private equity firm invests in customer acquisition, growth, or M&A activity to grow the organization, growth outpaces maturity and the system breaks down and leads to a number of problems.

**DevOps Patterns for Private Equity**

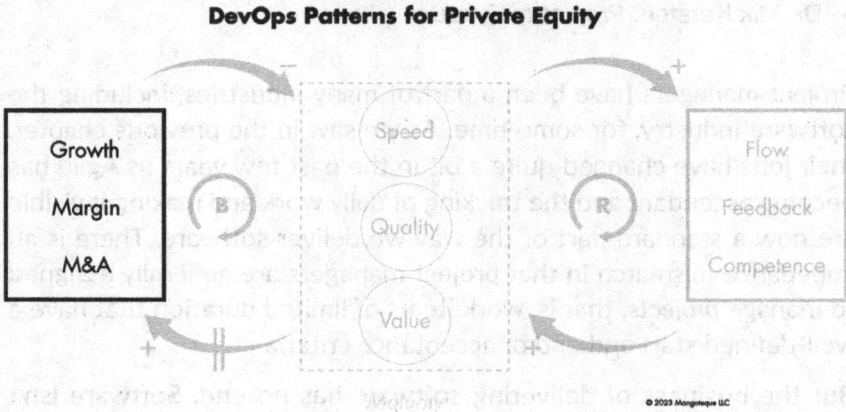

© 2023 Mangeteque LLC

# Lack of Improvement

Because the teams are continually trying to execute one-off tasks, they only have the ability to satisfy the current request and move on. Let's examine a hypothetical task from two different perspectives.

## Project Perspective

Allow the application servers access to a new database:

1. Create database

2. Determine access that will be needed and by what

3. Make a list of application servers

4. Create a playbook for modifying the security access

5. Execute the playbook against server 1

6. Execute the playbook against server 2

7. Execute the playbook against server...

This is tedious and repetitive toil. Because the goal is to finish the project, the next time this or something similar needs to be done, the same sequence is repeated from scratch. If someone takes the extra time to document the process, the next time would take less time. If they are trying to finish as fast as possible to move onto the next project, then the wheel will be reinvented again. There is no benefit from the previous round of work, and as the organization scales, the work scales linearly with it (i.e., toil).

## Product Perspective

Consider the alternative, when teams are assigned functional areas for which they're responsible.

Allow the application servers access to a new database:

1. Create database using a standard template so the database is configured for access in the same way as all previous databases.

2. Apply new access policy to all servers tagged as the necessary application servers by including them in the standard policy for the access to the new database.

In this case, the team is operating as if they have a product to support. They recognize that there is a lifecycle to infrastructure. The team creates and enables access to an infrastructure product. The difference is that for this type of product, the customers are internal, not those acquired by sales and marketing. In this situation, the investment of time isn't each time databases and access are needed, but up front, and the dividends pay off for years to come. Each time, the team will make small tweaks to make the process safer, easier, and more efficient so the cumulative effects can help the business to be more productive.

In a product mindset, we bring the work to the team, instead of the team to the work.

# All Aboard the Project Train

We had just formed a team to work on infrastructure as code. They were on their second week delivering as a team and had barely gotten familiar with Scrum mechanics. Almost immediately, requests for meetings started arriving. The security team wanted to have their input, the development team wanted to make sure their concerns were addressed. The team was still trying to figure out in which git repo to commit their first line of code.

What was happening?

The organization was used to delivering everything as a project. Iterative and incremental delivery was not in their vocabulary. They had learned that if a project was getting started, you needed to get your voice heard *early*, otherwise you might miss the train. The project would be completed and your concerns would not be addressed. Everyone wanted to get a meeting as soon as possible.

When we treat our deliverables as products, they can always be improved. Concerns can always be raised and the product will adapt, improve, and adjust. The process of improvement has no end date.

We assured those other teams that the Infrastructure-as-code team needed to get their house in order before attempting to honor outside distractions. Once they were in the flow, they would be happy to collaborate. And they were.

# Ops as a Cost Center

One problem with treating each task as a project instead of being part of a larger product is that often Ops is seen as a cost center, purely COGS. In this model, it's the developers who generate value through their R&D activity, and Ops is a drain to be minimized. Cost reduction instead of value creation. The faster and cheaper that Ops can be made to perform, the less drain.

However, Ops isn't purely a cost center. When operating with a product perspective, Ops is consulting, collaborating, and building tools that can be used by Dev and are a high-leverage point that enables and empowers Devs to build faster and with higher quality.

An engineering manager came to discuss the activity of the engineering teams over the last year. He figured that for the regular work the Devs had done, he'd allocate it as COGS, and any new activity would be R&D. For Ops, it would just be all COGS as usual.

"So, all the new ability to create things in the cloud, all the consultation on disk types, sizes, and allocations, all the testing done on the new storage system, all those things that Ops provided, none of it counts towards R&D?" I asked. "Hmmm," came the reply.

In a DevOps world, we don't throw things over the wall. The Dev and Ops teams collaborate to deliver business value. Goals are aligned, and collaboration yields outsized results.

## Finance-Driven Organizations

Perhaps because many portco CEOs used to be CFOs, private equity portfolio companies can often be very finance-driven. EBITDA is often a driving force behind how business decisions are made. However, EBITDA is an encapsulation of a number of business outcomes, and our focus here is on how to use technology organizations to *drive* business outcomes. That means that as the company cares about EBITDA, it also cares about on-time delivery of promises, attrition, customer retention, growth, etc.

A finance-driven technology organization will often adopt a project-based orientation. Often in finance, if I want to maximize the return on investment, then I want my investment to be maximally utilized. However, as we know from Agile: "You need to go slow to go fast." A pipe that leaks can often stop leaking if the pressure in the pipe is reduced. A group of engineers running at 100 percent will be *less* effective than one where there's slack in the system (recall WIP limits above). This is knowledge work, not machinery. Machinery parts that break can often be easily replaced, but humans who run at 100 percent burnout and have no ability to deal with variation, leading to attrition and missed deliverables.

This is a problem with shallow engineering metrics like lines of code (LOC), which are looking for output as opposed to value creation. In manufacturing, if a cog spins faster, we're likely to produce more product. In software delivery, more lines of code means *absolutely nothing*. I'll take an engineer who sits and thinks long and hard about a problem so they can write a short, easy-to-understand method of implementation that satisfies the business requirements, over one who quickly writes a long, complicated, brittle implementation that performs the same function.

The problem with fitting every budget request into a project that can be tracked and measured is that software isn't a project, it's a product. In his book *Project to Product*, Dr. Kersten describes the differences. With respect to budget, project-focused organizations perform "funding of milestones, pre-defined at project scoping. New budget requires creation of a new project." Product-oriented organizations, however, perform "funding of product value streams based on business results. New budget allocation based on demand. Incentive to deliver incremental results." Even without having read his book, you recognize these concepts from what we've discussed. Delivering incremental results!

Finance is poorly suited to understand the true costs involved in technology delivery. From their perspective, generally, the costs of software are the per-hour cost of labor, plus the costs of the server infrastructure to deliver that software (including source code repositories and testing). This is a very incomplete picture! There's no accounting for the costs of complexity, the cognitive load of one-off project after one-off project, the lost opportunity costs, or the nights and weekends people work beyond their salary expectations, which lead to burnout and attrition.

When those senior engineers finally burn out and depart the organization, they're often replaced with junior engineers because of market conditions. These engineers may have the same salary cost, but they certainly don't produce equivalent value. It's merely an artifact of market pricing. The office of the CFO needs to work closely with the office of the CTO, but it should not drive the latter's organization.

# *Costs*

Often Finance only sees revenue and some bills. Below is an incomplete list of costs for maintaining marginal products that are not easily measurable in dollars:

- The delay in adding the new hire to the rotation because they have to learn the oddball system.

- The cognitive load of being on-call for something that is barely tested or maintained.

- The activity to patch and deploy something that is hard to build and scarcely tested.

- The periodic fires that need to be put out when a customer does something unexpected or when the company is the subject of a denial-of-service attack.

- The cognitive load on the security folks for a product with known XSS or CSRF vulnerabilities that they know no one will patch.

- The time it takes to run security scans and tests just so they're documented, not because they'll be fixed.

- The additional burden on the one SRE who still understands something about the marginal system because they've been around the longest.

- The cost to bring in a developer contractor to perform one-off fixes on the legacy product because it's been "unstable lately."

- The load on the project manager who still has to report on the status of the marginal product because, if she doesn't, people might forget that it exists.

- The additional attack surface the company is vulnerable to for a product that barely makes a profit.

- The time Customer Success spends on calls for the marginal product instead of developing stronger relationships with high-value customers.

- The context-switching staff must do when paged about the oddball product, which takes them away from high-value work.

- The toil (deleting files, growing partitions, adding resources) that must be performed to keep the site up and running because no software runs forever without intervention.

- The multiple contracts (licenses, service contracts) that must be maintained for legacy software tools that are needed to support the legacy product.

- The exceptions that must be carved out in new platforms and processes to account for a system that won't be modified because of the difficulty in doing so.

- The different meaning of terms used in the marginal product because it was either brought in via acquisition or named before the rest of the company had standardized.

- The additional difficulty it takes to troubleshoot problems with the product because it doesn't use modern or standard company tooling, and everyone who developed it has left.

- The efforts, contracts, and manual work it takes to maintain this one product in a data center when everything else has moved to the cloud because "our monthly data center bill is cheap!" If it stays within its power and cooling allotment, of course.

- The stress on the VP of Engineering when they hear that Shonda has been interviewing and she's the only one who still understands that system.

- The promise Engineering gave to the company that they would sunset the product in three years, but three years later, they're finally beginning to make progress, and it will only take three years (from now).

Engineering work being driven by Finance is akin to the reviews by change advisory boards — work being reviewed by people who have little understanding of the thing they're reviewing. I recall a client engineering manager who was a half-million dollars under budget. Finance asked her to justify why she was hiring a contractor at $37/hour instead of the $34/hour that had been her original *estimate* of what it would cost. Her response was, "I'm a half-million dollars under budget right now. If you're going to scrutinize every dollar I spend, what's the purpose of giving me a budget?!"

When technologists become accustomed to having every dollar they spend scrutinized, they learn to avoid spending entirely. In his essay, "5 B's of Budget,"[20] Ryan Parker of Mercato Partners cites a major pitfall for CFOs: cost reduction at the expense of value creation. I had a client working on a major automation initiative that would eliminate an entire class of work for approximately eight engineers; about half their

---

20  https://mercatopartners.com/resources/5-bs-of-budget/

workload. To test the software that would be needed for this initiative they would need to spend approximately $300/month for virtual machines. They were afraid Finance would question the expenditure. The project almost stalled because these folks were afraid to spend $300 to save the equivalent of four engineering salaries. There's no question that the CFO would have been thrilled to spend $300 to unlock the value creation capabilities of four engineers.

Aside from value creation, as we will see, the biggest cost reductions are impossible to achieve by watching every dollar that's spent on a project. That may move the needle a few points. Much larger cost savings can only be achieved through engineering.

# Undifferentiated Heavy Lifting

There's a lot of muck involved in running a company, especially any company that delivers value through software. There are ERP systems, CRMs, source code repositories, documentation systems, monitoring, testing. Each of these systems needs to have an owner, hopefully in the form of a team, that's responsible for all the things necessary to keep them available, up to date, and secure. If it's everyone's responsibility, it's no one's responsibility.

When Jeff Bezos first launched Amazon Web Services, he framed the cloud computing platform as one that could do the undifferentiated heavy lifting. That is, all the things that are necessary to run infrastructure that are no different from the work that any competitors have to do. Essentially, all the things that *everyone* has to do.

In the DevOps movement, because we're focused on delivering value to our customers, this muck is the type of work that doesn't deliver differentiated business value. I often tell clients that if their ability to run a source control repository is going to be the thing that separates them from their competition, then they're in the wrong business and should become a source code repository company.

Examples of undifferentiated heavy lifting services that can be run by others:

- Revision control systems
- Continuous integration/continuous delivery (CI/CD) tools
- Ticketing systems
- Observability/monitoring systems
- Artifact repositories
- Feature flagging tools
- Standard databases
- Incident response tools
- Security scanners
- Exception handler trackers

All this undifferentiated, heavy-lifting-type work should be farmed out to companies that are solely focused on delivering that service as their own way of providing business value. Industry luminary Adrian Cockroft has said that a monitoring system needs to be more highly available than the systems that it's monitoring. Should a portfolio company spend more engineering time on its monitoring system (99.999% uptime) than the product that their customers pay for (99.9% uptime)? Of course not.

Managing work through an organization and delivering heavy lifting capabilities (internal developer platforms, resilience, scaling mechanisms) to internal customers as a product is difficult enough. It makes no sense to build and maintain a product when the competition is getting a world-class version of that same service by using a credit card.

## Offshoring

But what if we want to do differentiated heavy lifting? What if we want to rewrite a monolith into microservices or add product functionality or create a unit test suite for a product that has none. Many portfolio companies supplement or replace parts of their staff with outside vendors or hire directly, either short term or long term, to get these benefits.

Moving work offshore can give access to a global talent pool. Offshore staff is distinct from outsourced staff. Back in the bad old days, Engineering used to throw things over the wall and Ops would have to deal with it. Then came DevOps to fix the problems that come from this model. The lessons of DevOps tells us that if we set up our organization so that our outsourcing partners are someone we throw things over the wall to, then we're repeating the same problems we've committed to eliminating.

Unless this is done carefully, we can put ourselves in a worse situation than where we began. If the outsourced team hands its work off to the existing staff, now that existing staff has even more things they're responsible for without knowing how any of it was built! If we do temporary staff augmentation, we must ensure that the remaining staff are part of the teams that are building the new thing.

If possible, in these situations it's best to offshore the work, by which we can supplement our existing teams with staff from lower-cost geographies that will be around to maintain what is built. An operating advisor explained that they would gladly pay US$40K/yr for offshore staff that would cost $25K/yr to hire themselves because of the ability to swap employees until they find the ones who are the right fit. These are not just engineers to do the work that no one else wants to do.

Another reason would be to create a follow-the-sun model where the value stream can be continually delivered around the clock, and in the event of an outage, no one needs to be woken from their slumber.

For these situations, it's important to remember two things:

1. Our default mode should always be to build robust systems that handle failures automatically. We shouldn't hire people in other parts of the world to do work that a computer should be doing.

2. Make sure you have the local management talent in place to support those teams. We neither want engineers without a manager they can turn to for help, nor managers who have to work odd hours to support their team at the expense of their personal lives or family obligations.

Other challenges may arise if our product deals with government or healthcare. There may be regulations about who has access to different environments. If we're doing infrastructure as code, for example, the same code is used to build *every* environment, so there is still a position of high leverage.

The biggest challenge to these teams (and any distributed team) is time zones. For this reason I generally recommend "nearshoring," finding talent in a closer time zone. For North American companies, this generally means Central or South America. For European companies, this often means looking at Eastern Europe. There doesn't need to be perfect overlap, but for engineers to be able to collaborate with other members of their team during the regular workday, instead of off-hours, is invaluable.

My experience is that offshoring (or better yet, nearshoring) can give access to incredible talent that has a lower cost of living than in other areas. Some of the most talented engineers I know are from all different parts of the world. Build these teams with inclusion and support, and they can yield great results for your portfolio company.

# Burnout

Of course, all the effort to manage a never-ending stream of project work, as well as spending time on work that can be purchased by a credit card, is not without consequence. People don't have infinite capacity to work we do not have unlimited engineering tokens, and people do not do their best work when they're tired and will eventually burn out.

We did a service delivery assessment for one company where they had a single engineer on-call 24/7, 365 days a year. We wrote in the report that they must add more people to the on-call rotation because the engineer was *already* burned out. During the follow up with that company a year later, we asked how he was doing. "We didn't get him help and eventually he had a nervous breakdown and had to leave." It was plainly unethical and should never have happened. Additionally, they lost one of their foremost technical experts and the only person protecting the revenue stream of that acquired company! This was a poor decision on many levels.

I experienced one of the hardest lessons I learned as a growing engineering leader when a senior engineer in one of my companies turned in his resignation. We'd been building a new platform for the company and were mere weeks away from it being fully rolled out. He'd been on-call 24/7 for months at the same time as one other engineer. I had repeatedly urged, suggested, and cajoled the two of them to create a rotation until we were able to fill an open role for another engineer. After enough months, he'd had enough and quit. If given the opportunity again, I would have created the rotation myself without any repeated suggestions.

Too many times when private equity combines multiple companies, chaos is allowed to reign for too long. Because things are mostly working, leadership will allow wildly different ways of doing deployments and skeleton teams are left to support existing products, and then because the business wants growth those teams are given more work. As the organization, traffic, and demands grow without the corresponding necessary maturity, things begin to bog down. I'm often called after a significant portion of the senior staff has quit in frustration.

**DevOps Patterns for Private Equity**

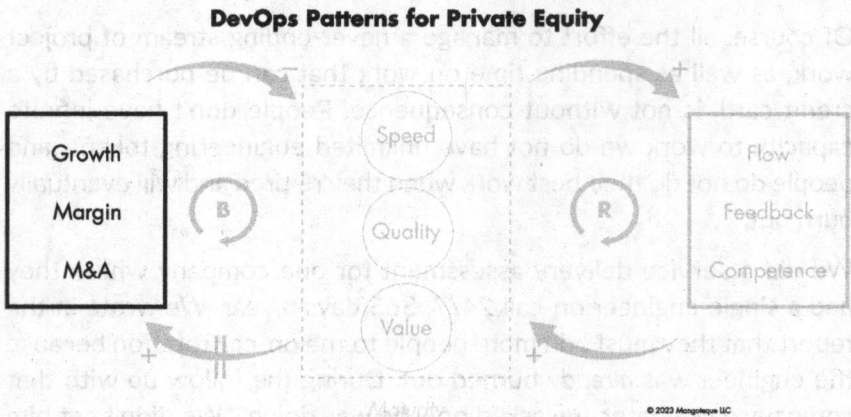

Thankfully the DevOps movement has taken the topic of burnout very seriously, embracing the work of experts like social psychologist Dr.

Christina Maslach. Bringing in experts from other fields to learn their lessons is a strength of the DevOps movement.[21]

# Connecting Leaders

We've decided that we want to change to a product orientation. We're still deciding what the new team structures will look like. We want to watch out for burnout on the engineering teams. What will these "product teams" be? Who will staff them? How will we get people to buy into the changes?

Frontline managers to the rescue. In engagements, these are the folks I work with most often. In order to improve the flow of work and the work itself, we need to *understand* the work. No leaders in the organization are closer to the work than the frontline managers.

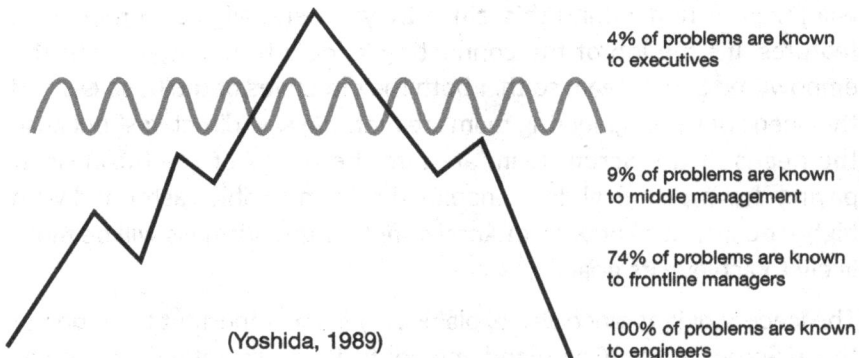

4% of problems are known to executives

9% of problems are known to middle management

74% of problems are known to frontline managers

(Yoshida, 1989)

100% of problems are known to engineers

The Iceberg of Ignorance. As the story goes, in 1989, Sidney Yoshida published a paper about an auto manufacturer describing awareness of problems at different levels of the organization. Whether the numbers are entirely accurate or not, the lessons of the "Iceberg of Ignorance" are instructive.

If we want to understand the data about the nature of the work in our organizations, we have to go to where the data is. In her *Harvard Business Review* essay "The Real Value of Middle Managers,"[22] Dr. Zahira Jaser calls these managers who are closest to the information the frontline managers, *connecting leaders*.

........................................................................................

21  https://videos.itrevolution.com/watch/524022536/
22  https://hbr.org/2021/06/the-real-value-of-middle-managers

*"This concept recognizes that every leader is also a follower, and every follower is also a leader. Thus, a manager in the middle of hierarchical layers builds relationships with those at the top (from a position of followership and lower power) and with the people at the bottom (from a position of leadership and higher power)."*

— Zahira Jaser

Depending on the size of the organization and its structure, there may be more or fewer layers of middle managers. Regardless of the structure, the managers who are closest to the work are the leaders with the most information about the work itself. These connecting leaders understand and communicate the needs of the business to the engineers, and also understand the needs of the engineers. If an engineering team needs time to eliminate a serious piece of technical debt, there is almost no chance that folks in the product organization will know or understand this, since they're generally concerned with features. It's the job of the connecting leaders to negotiate with the empowered Scrum team so that both the directives of the business and the needs of the engineering team are met. We say "directives" because the needs of the Scrum team are *also* the needs of the business. If paying down technical debt enables the team to ship faster and with higher quality, we know from *Accelerate* that the business will be more likely to achieve its goals.

The iceberg of ignorance also explains why it's so important to empower those Scrum teams. Command and control is more aptly described as command and chaos since it's impossible for those at the highest level of authority to understand the local decision making at the edges. This isn't a failure in the leaders, it's human nature (and biology is messy). It's impossible to consume the volume of detail at the furthest edges of our sphere of influence.

Some of the most command and control organizations in the world have recognized this and allow autonomy at the edges. Modern Western militaries develop a strong NCO (non-commissioned officer) component of their organizations because objectives formulated at headquarters need to be implemented on the battlefield. The NCOs allow them to do just that by synthesizing the objectives and the information on the ground at the time. In his book *Turn the Ship Around*, Captain David Marquet (Ret.) talks about how he had active, engaged, knowledgeable sailors on his nuclear submarine, and they were able to achieve the highest ratings ever given by the U.S. Navy, while other submarines continued to rely on the captain to be the smartest person in the room. Talk about asymmetric warfare!

When we want to make change in our organizations, we need to bring decision making to where the information is. This is the best way to make the highest-quality decisions. Every organization loves to talk about how they're "data driven." Bringing organizations around to a DevOps mindset is a *cultural change*, not a technological one. When you want to make a technology change, send in the architects to get their hands dirty. When you want to change the way teams operate, from a project to product mindset, send in the connecting leaders.

*"If we have data, let's look at the data. If all we have are opinions, let's go with mine."*
— Jim Barksdale, CEO, Netscape

## Westrum Generative Culture

The DevOps movement has talked for years about generative culture as described by Professor Ron Westrum in his paper *A typology of organizational cultures.*[23] It's featured prominently in both the *2019 Accelerate State of DevOps Report* and the *Accelerate* book as a way to describe and categorize the desired organizational culture necessary to be an elite performer. In recent years, Westrum has been featured at DevOps conferences, interviews, and podcasts.[24]

In his paper, Westrum discusses medical units, but we understand that his classification system is broadly applicable to local governments and businesses. He describes three types of organizational structures: pathological, bureaucratic, and generative, shown in Table 1.

23  https://www.ncbi.nlm.nih.gov/pmc/articles/PMC1765804/pdf/v013p0ii22.pdf
24  https://itrevolution.com/podcast/the-idealcast-episode-17/

| How organisations process information | | |
|---|---|---|
| **Pathological** | **Bureaucratic** | **Generative** |
| *Power oriented* | *Rule orientated* | *Performance orientated* |
| Low cooperation | Modest cooperation | High cooperation |
| Messengers shot | Messengers neglected | Messengers trained |
| Responsibilities shirked | Narrow responsibilities | Risks are shared |
| Bridging discouraged | Bridging tolerated | Bridging encouraged |
| Failure > scapegoating | Failure > justice | Failure > inquiry |
| Novelty crushed | Novelty > problems | Novelty implemented |

Table 1

I've used this typology with many portfolio companies to demonstrate where I feel they fall on the scale for each of the three process types. People often recognize their own organization in certain columns with horror.

Westrum instructs that, "The scheme concentrates on information flow as a key variable." Astute readers realize these types of schemes are created by leadership and cannot be created by the engineers. Consider our example with Captain Marquet above. His submarine was effective because he encouraged collaboration and he distributed responsibility so that risks were shared, the sailors collectively learned from failures, and they were always looking to improve (novelty above). Westrum explains: "Information flow is higher in a generative situation, where managers see themselves as coaches rather than commanders."

Generative organizations focus on three key outcomes that result from this optimized flow of information. They almost perfectly align with Gene Kim's three ways of DevOps:

- Alignment — teams work collaboratively to achieve common business goals

- Awareness — we make work visible

- Empowerment — we enable teams to have a say in the work they'll perform because they're closest to the information

**DevOps Patterns for Private Equity**

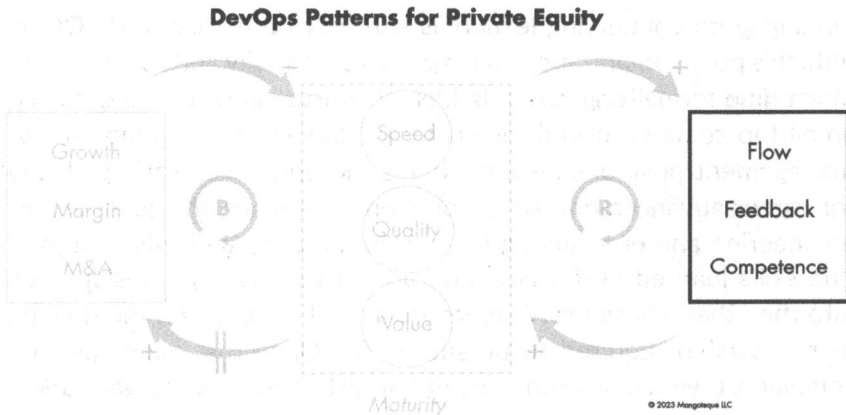

We talk in incident response about the need for blamelessness in our learning reviews. Why? If messengers are "shot" or failures lead to scapegoating, then people will simply suppress the information. No one wants to lose their job or be made to feel bad. The information that could be used to prevent the outage from recurring, or reduce its duration, will be hidden. There's no opportunity for continual improvement in an organization where novelty is crushed. These organizations will have lower submarine ratings than those that embrace a generative culture, or they will be crushed by their competition in business.

*"A generative culture will make the best use of its assets, a pathological one will not."*
— Dr. Ron Westrum

## Moving from Tactical to Strategic

Often it's difficult to find the connecting leaders who are necessary to create the changes in the organization. When multiple organizations have been combined, as with platform companies and tuck-ins, there are different levels of senior engineers, principal engineers, senior managers, etc. in the different organizations. Now they're one company and the hierarchies need to be normalized, including pay scales and titles.

Finding good connecting leaders is hard but necessary work. Often, until this point, organizations have grown organically and haven't spent much time formalizing career ladders. Unfortunately for our industry, an all too common occurrence is for engineers to be promoted into management positions because that's the only recognized pathway for career advancement. As anyone who has made this leap knows, engineering and engineering leadership are *completely different jobs*. The skills learned performing one do not necessarily translate well into the other. These new leaders are almost never given the training or mentorship required to be successful. Once the companies are combined, they can be even more out of sorts in the larger organization.

When making my own transition from engineering to leadership, one of the things I struggled with most was the question: What is work? As an engineer, my job was to build things. Simple. If I spent time in meetings, even if those meetings were necessary and useful, it was time that took away from my building things.

Once I became a leader, I spent the *majority* of my time in meetings. My engineer brain kept telling me that I wasn't getting anything done! Only after understanding that going to meetings was how I actually delivered work, by discussing needs, making decisions, and setting engineers up for success, did I overcome the cognitive dissonance resulting from so many years of building things as a way of delivering work. Strategic work is work! No less difficult than tactical work, but capable of much greater impact. Instead of writing the code, I could enable many people to write great code and achieve the goals of the business.

## Career Ladders

One of the great things about the Salesforce organization was the well-defined paths for both engineering and management. It was very clear what the next step was on each career path and even how the different paths mapped to one another. For example, as Engineering Architect, I mapped equivalently to Senior Director in the management track. This was a 16,000-person (at that time) organization, even if there were maybe 100 architects.

When it came time to develop a career ladder for my own engineering organization, I made sure to make clear to the engineers that with each step up, there was an increased scope of responsibility. There would be larger scope to the projects in which they were involved, and they would be expected to not only understand the details of specific components but how they worked with and fit into the larger whole.

I have often recommended Duo Security's career ladder to folks as an excellent example of how it should be done.

## Ladder

| Level | I | II | III | IV (Senior) | V (Principal) | VI (Architect) |
|---|---|---|---|---|---|---|
| **Experience, Scale, Scope, Skill** | | | | | | |
| Typical Years of Experience | 0–2 | 1–2 | 2–4 | 5+ | 10+ | 15+ |
| Scope of Influence | Self | Project | Team | Engineering | Product Development Organization | Duo |
| Time Horizon of Effort | 1 sprint | 2 sprints | 2+ sprints | 1 quarter | 6 months–1 year | 1–2 year |
| Practicing Skill Level | Beginner – Demonstrates a limited use of a competency, requires additional training to be able to apply without assistance or direction. | Intermediate – Demonstrates a functioning working competency and is effective without the need for direct support. | Advanced – Demonstrates an in-depth application of competency and is able to consult or lead others in their application of it. | Expert – Demonstrates a mastery of competency and is recognized as an authority on the subject by peers and leaders. | | Master – Evolves our collective understanding of the subject / practice. |
| **Technical** | | | | | | |
| Summary | Responsible for implementing and maintaining scoped product or | Responsible for scoping and implementing individual-level | Responsible for designing and implementing team-level | Responsible for designing and owning technical solutions for | Responsible for solving complex, mission-critical problems with an | Responsible to company for complex mission-critical |

SOURCE: Erik Barbara (2018), "Engineering the Business: Duo's Software Engineering Ladder," https://duo.com/blog/engineering-the-business-duos-software-engineering-ladder.

The summary line leaps off the page with the repetition of "Responsible for ..." Each increasing level moves responsibility from the individual level to the team level and beyond.

Having a well-defined career ladder enables engineers to understand how they fit into an organization, even after a number of acquisitions. It helps them find their place on a map. In every company trying to move up to the next level, having leaders with large areas of responsibility buy into the new ways of working, new ways of managing work, and the new path forward, is critical for success.

# Making Change

*"I remind the reader that the improvements took place with the same people and no new equipment."*
— W. Edwards Deming, *Out of the Crisis*

No two portfolio companies are the same. Each is at its own stage of its own journey. Sometimes it's only a few companies that have been merged, sometimes it's many. In any case, none are good and none are bad, they are simply where they are.

Being able to make the right changes for each situation is often what separates the good private equity firms from the great. As Toyota recognized, they can all improve. Choosing what type of change to make in the technology organization is rarely easy. Stumbles are a part of the process, and people need to have faith to not fall back.

This is strategic change, and as we've discussed, it cannot come from the front-line engineers. Leadership (technical or management) needs to put the conditions in place to make change happen. As Microsoft engineer Bridget Kromhout teaches, containers will not fix your broken culture.[25] In his books and lectures on human factors in safety, Professor Sidney Decker argues strongly against the "bad apples" theory. People don't

---

25  https://queue.acm.org/detail.cfm?id=3185224

cause aviation accidents through carelessness when their own lives are the ones on the line. Engineers labeled as bad apples are often seen that way because the system is "working as intended." Often they're labeled as the problem with the idea that putting another engineer in the exact same situation will yield different results. I've talked to portfolio companies where multiple engineers in one department were put on performance improvement plans or reassigned. I became suspicious.

I was talking with a senior SRE (Site Reliability Engineer) at a portfolio company and he was frustrated. He was working nights and weekends and was still being told by his boss that his development team customers weren't satisfied.

"How many teams do you support?" I asked.

"Four," came the reply.

"And you go to standup for all four?"

"Yes."

"OK," I said. "Let me guess. You get work from all four teams and it's up to you to decide which one is the most important for you to work on first, and they don't talk to one other."

"Yes, exactly."

"So, no matter which team's work you choose, the other three teams are guaranteed to be disappointed."

"Yes, exactly."

The system was set up for him to fail 100 percent of the time.

Because each company is different, there are almost an infinite number of paths that can be taken. In each case, the goal is to create a learning organization that is always improving. Sometimes even the skeleton isn't there and it needs to be created. Sometimes, some attempts at reorganization have already happened, but there's no shared vision.

## Structural Transformation

*"They operate like 47 lone wolves."*
— Engineering manager

I've written that process is a sign of success.[26] When companies have been acquired by private equity, they've done something right! Now, as they move to the next level, they often need to think carefully about process.

The organization begins small, with many scrappy engineers doing what it takes to get the job done. As the company grows, there's more work, and they add more staff to handle the increased workload. Over time, different people become experts on different types of work, but the incoming flow of work is always increasing.

---

26  https://blog.mangoteque.com/blog/2021/02/09/process-is-a-sign-of-success/

**DevOps Patterns for Private Equity**

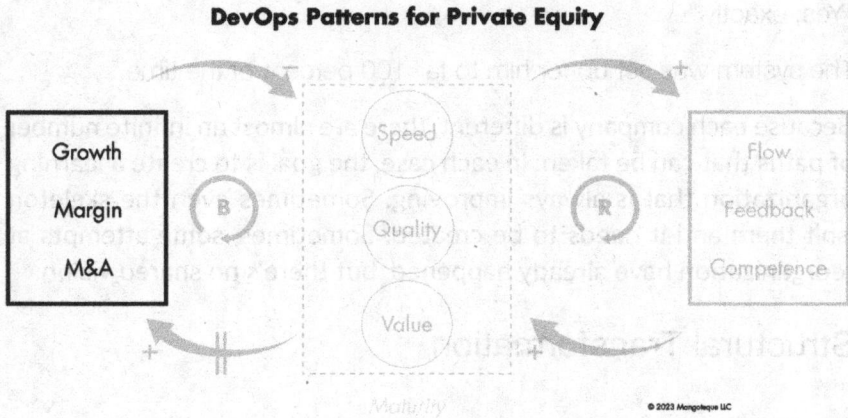

Growth

Margin

M&A

Speed

Quality

Value

Flow

Feedback

Competence

© 2023 Mangoteque LLC

Eventually, systems are put in place so that work no longer comes in via various channels (e.g., email, chat, drive by) but through a centralized ticketing system. The developers usually split into multiple Scrum teams (stream-aligned teams) dedicated to specific products or features, but the operations staff becomes buried under an ever-increasing flow of work that eventually overwhelms them. The developers are unhappy that everything takes too long, and they're frustrated. Management becomes frustrated, and leadership calls for help.

Organizations in this spiral are struggling under their own weight. Continual improvement is almost impossible when these improvements are happening on an individual basis, like a collection of lone wolves. As we discussed earlier, each of these individuals working in isolation is taking a *project*-based approach to the work. Each engineer gets to reinvent the same wheel. As the work comes in, there's a project of some size that's executed to get the work done. Sometimes there are even project managers assigned to run around the organization and collect the status of each of these individual projects so they can be reported to management in the latest presentation style (RACI and the like), generally in watermelon charts (green on the outside, red on the inside).

With these organizations, it's important for operations leadership to sit down and discuss changing the structure (i.e., the system) of the organization into *product* teams. Instead of 47 lone wolves, there are approximately six or seven two-pizza teams. (This is the Amazon rule that a team can be fed with no more or less than two pizzas. Above this number, people tend to naturally start to split into two teams based on like concerns.) These teams can be divided by services (e.g., front end, back end, database) or by functional areas (e.g., resilience, observability, platform engineering, developer tools). In each case, the new teams work together to deliver a product; one that continually gets better, is improved, is iterated upon, and provides self-service tools to the engineering staff, enabling them to "shift left."

As a client used to say: "the best service is no service," meaning that when teams are enabled to service themselves, there isn't an ever-increasing amount of work being funneled into operations. These internal product teams are continually making the hard things easy, so that they can play an active role in helping the business grow and meet its objectives. No longer is the flow of work always increasing, no longer are there repeated calls to hire more staff, whether onshore or offshore. Instead, the team accomplishes more with the same.

## Vision (Treading Water)

In other organizations, the divide into two-pizza teams has already taken place. Sometimes there will be an SRE team and a developer tools team. Sometimes there's a team responsible for databases. In many respects, however, these teams are just 47 lone wolf teams in miniature. That is, instead of incoming work able to go in 47 different directions, it can go in four or five.

What's the difference? What's missing? Typically, when teams are still trying to work faster even though the work is now divided, there is no vision. There is no direction. I often ask: What are you trying to do? Who do you want to be?

These teams don't meet with each other. The same work, accomplished differently, could happen on multiple teams without anyone being aware. There is no overall idea of what the end state looks like, there is just more doing, just like the wolves. The analogy I've used to describe this is treading water. The teams are treading trying to keep their heads above water. As more work comes in, they try to tread water faster. Instead, they need to swim. They need to pick a direction and go. Even going in the wrong direction is an opportunity for learning and course adjustment. The reason the teams are still working independently is that there is no shared vision of who they are and where they want to go. Decisions about technologies and processes are made independently.

For example, if a team is working on CI/CD systems, there's little discussion with the Quality Engineering or Security departments about which tests will run as part of the pipeline and how those tests will be run. There's little discussion with the Infrastructure teams about how those pipelines will be specified, or built, or, ideally, self-provisioned.

With these organizations (and with most organizations!) I recommend a weekly architecture meeting — time for engineers to come together to discuss projects they're planning or work they're struggling with. It gives them an opportunity to use the collective intelligence and experience of the group and collaborate. This collaboration can happen within the Ops teams or with other groups they bring in as well (like Quality Engineering). It also provides a safe space for junior engineers to develop presentation skills, for senior engineers to teach how they approach these problems, to agree on standards for the organization, and to decide on next steps for determining what will be the most successful paths forward.

Why don't we just discuss these topics at the daily standup during the parking lot? Discussing during the parking lot means that the 15-minute standup (where we *stand up* to keep it short) becomes an hour-long (or more) discussion with a group of people who aren't prepared to have the discussion to begin with. This leads to arguments, half-baked ideas, and calls to do more research. It keeps others who will be affected or who have valuable input from being involved in the discussion. The Jeff Bezos six-page memo is a great example.[27] Each meeting starts off with people being prepared for the discussion that is to come. This is not achieved by one-off conversations at standup.

---

27  https://www.cnbc.com/2018/04/23/what-jeff-bezos-learned-from-requiring-6-page-memos-at-amazon.html

The output of an architecture meeting is often a Technical Specification that can facilitate what engineers love: to build. (See Appendix A: Document Templates)

# An Agile SRE Meeting Plan

Engineers dislike meetings. What engineers really dislike are meetings for which they perceive no value. Below is described a meeting plan developed, iterated upon, and used over many years at multiple companies that has proven very effective at both maximizing meeting value and minimizing unnecessary time in meetings, so that engineers may do what many enjoy most: building things. This may not be the perfect plan for your organization, but it will hopefully inspire conversation and discussion about how to structure the time of *your* SRE team.

## The Structure

Over the years, I've experienced many different Agile implementations. Scrum is considered to be a pretty poor match for interrupt-driven teams like Site Reliability Engineers (SREs), but how do we get the agile benefits of Kanban and still retain many of the advantages of Scrum? How do we have a schedule that's relatively light on meetings, but still keep the maximum amount of communication and transparency? How do we continue to be *agile* instead of just *doing* Agile, especially with distributed teams?

In the plan outlined herein, we try to balance many of those things. We lay them out as Monday to Friday, but they could certainly be Tuesday to Tuesday, or whatever fits best to line up with the development team's sprints. (Pro tip: line up the best you can with the development team sprints.) When we first embarked on this path, we were on two-week iterations, to match what the Dev teams were doing. Over time, we discovered that we lacked the responsiveness we wanted to provide to those teams, and thus switched over to one-week iterations in order to maximize (internal) customer satisfaction.

| MONDAY | TUESDAY | WEDNESDAY | THURSDAY | FRIDAY |
|---|---|---|---|---|
| ITERATION PLANNING | DAILY STANDUP<br><br>INTER TEAM SYNC<br><br>ARCHITECTURE MEETING | DAILY STANDUP | DAILY STANDUP | DEMOS/ RETRO |

The iteration starts off a bit meeting-heavy to emphasize alignment, and then allows for plenty of time for our standard SRE work (elimination of toil, etc.), and finally closes the iteration with time to reflect and improve.

## Iteration Planning Meeting

### Purpose

The iteration planning meeting is, well, just what it sounds like: planning the iteration. Because SRE teams can be interrupt-driven and use Kanban, iteration planning is not for committing to delivering specific work in a time box (like Scrum). Instead, it's for making sure that the entire team is on the same page in terms of priorities, needs of the business, project work assignment (e.g., Anne really wants to work on the VPC network project), dependencies on other teams, urgency of different tasks, and looking forward to the next few weeks and what work may be taken on.

This is really a time for discussion, and for identifying things that will require more in-depth discussion. It's not a time for going deep on any particular task, but a time to make sure that everyone on the team is aligned for the next batch of work. Oftentimes, work from the previous iteration will simply continue in the current one, but this is also a good time to check that the work is being delivered to expectations, especially as a result of the demos (which we'll get to later) that closed out the previous iteration.

Because we want to *be* agile, instead of simply *doing* Agile, work that has been agreed to *isn't* set in stone for a week. We just don't want to oscillate wildly from day to day, or even week to week, and the iteration planning meeting is the opportunity to ensure that the team is moving in the same direction simultaneously. Because we're "walking the wall" while negotiating tasks, this is also a great time to recognize blocked work and any of the "time thieves" Kanban expert Dominica Degrandis describes in her book *Making Work Visible*.

At the end of this meeting, the team leader should have worked out with the team a balance of work being requested by other parts of the business and work proposed by the team itself. In some iterations other teams' requests will occupy more time, in some iterations the needs of the team will take priority. A successful manager will navigate the balance between the two, ensuring that the needs of the business are being met while simultaneously allowing the team to reduce toil, perform chaos engineering experiments, collaborate with other teams, etc.

## Mechanics

The iteration planning meeting should begin with an already prioritized Kanban board. The team can negotiate changes to the priorities during the meeting, but this is not a time for debating what it is that the business values. Depending on the size of the team, and the amount of discussion needed for specific tasks, this meeting should take no more than one hour. Past this point, engineers will lose focus and interest and "just want to start getting things done." The meeting should include:

1. Holdover discussions from the previous iteration

2. Explanation of the top priorities from the business

3. Explanation of the top priorities from the team

4. Identification of merged prioritization

5. Identification of resources working or interested in tasks

6. Assignment of work if there are no resource volunteers for necessary work items

7. Parking lot

# Daily Standup Meeting

## Purpose

The daily standup has the same purpose as it does in Scrum — to keep the team in close alignment with respect to deliverables and to identify any items that require assistance from other team members or management in order to keep the team operating at its highest velocity.

This plan only has standup during the middle of the iteration because the beginning and end already have time for team discussion at the iteration planning and iteration review meetings.

The daily standup should be restricted to the work at hand and should not devolve into in-depth discussions on specific tasks. These will extend the meeting and hold the rest of the team hostage for the duration of the discussion.

## Mechanics

1. What did I do yesterday, what am I working on currently, and identification of blockers — for each team member

2. Parking lot

If working with distributed teams, you might want to allow the standup to extend to a full 30 minutes to allow the teams to socialize with one another and therefore build some of the bonds you would otherwise get by colocation. In that case, the standup should conclude as soon as the parking lot is complete.

# Inter-Team Sync Meeting

*Purpose*

If good intra-team communication is difficult to do well, then good inter-team communication can be even more difficult. Couple this with dependencies between teams, and you can easily see the need to set aside an agenda specifically for this purpose.

The inter-team sync is to ensure close coordination and transparency between the SRE team and their *primary* customer. We don't want to fill each iteration with sync meetings between the SRE team and *any* customer they may have because that number may be very large, and frequent context switching is a major impediment to delivery of work. But, the team that works most closely with the SRE team should have a short meeting to discuss work in progress, dependencies, upcoming projects, etc. In this way, we attempt to ensure that both teams are working at the maximum safe velocity and minimize misunderstandings and the conflicting priorities and unknown-dependencies time thieves (see Degrandis above).

The inter-team sync is how DevOps is done!

*Mechanics*

The inter-team sync meeting should be no longer than 30 minutes. Any discussions of deep architectural questions should be put on the agenda for the architecture meeting. There should be an agenda (we like Kanban boards for this purpose), created and maintained by the team leads or managers, that's widely available, that anyone can contribute to, and that tracks all the work items shared between the two teams, especially dependencies.

The person who runs the meeting simply "walks the wall" until there are no more items to sync and then ends the meeting.

# Architecture "Arch" Meeting

*Purpose*

If the iteration planning meeting and the daily standup are *not* the place for in-depth discussion, then the weekly arch meeting is *exactly* the place for such discussion. This is the forum for any deep technical discussions on the SRE team. This is also a forum where members from other teams can either be invited or be regular attendees to give guidance, ask questions, provide clarification, etc., of work with which the SRE team is tasked. In other words, DevOps!

The outcomes of inputs to the arch meeting are often technical specifications, diagrams, documentation, requirements documents, and experiments. This can be a time for senior staff to give feedback on proposals to other, both senior and junior, members of the team. This can be a time to solicit opinions from the group on a new or existing technology or to review past postmortems. This can be a time for helping to figure out how to navigate toward a long-term goal. The opportunities are wide open (by design), but the goal should be that by the end of each arch meeting, the entire team should have taken a step toward achieving the goals of the team and of the business.

The number of times I've heard the phrase "let's table this and add it to the agenda for the arch meeting" over the years are far too numerous to be counted. This is another opportunity for the team to ensure that they're highly aligned as they move into the meat of the iteration.

## Mechanics

The agenda for the architecture meeting tends to build itself over the course of the previous iterations. We always use a simple Kanban board or Google Doc for keeping track of proposed topics. The person running the meeting can cover each topic in turn, or it can be run Lean Coffee style,[28] or, if someone has an especially important topic for discussion, that can be moved to the beginning or end of the meeting or the end (to allow more time for open-ended discussion). It's really up to the attendees to determine which best suits the style of the teams involved.

---

28 https://agilecoffee.com/leancoffee/

# Demo/Retro Meeting

## Purpose

Students of Gene Kim's three ways of DevOps know that the second way is all about feedback. In order for this to be successful, we need to set aside time in our week (in the form of a meeting) to specifically enable that feedback to occur. The demo/retro meeting has two purposes:

1. To have the team demonstrate (demo) the work they accomplished (not necessarily completed) during the iteration.

2. To have a retrospective (retro) to discuss how to improve the team in a *psychologically safe environment.*[29]

The demo allows the team to get fast feedback on work they have completed or that's already in progress. There's a saying in Agile, "maximize the work not done," which reminds us to spend our time on work that's critical to our success. If someone is delivering a project that doesn't meet our needs, we'd like to give them that feedback *before* they finish the project, not after, so they can adjust course. The bar for a demo is extremely low; unit tests, working demos, command line tools, a single API call are all acceptable demos. The point isn't to dazzle, the point is to demonstrate working code.

The retro gives the team the space to improve in the Kaizen fashion. We follow the traditional retrospective format (what went well, what didn't go well, what could be better) with some modifications. The goal is for each team to be higher performing at the end of every year than they were at its start. By setting aside a safe space for the team to talk about how the iteration went for them, and how the team can improve, we're creating an environment that fosters and encourages that improvement.

## Mechanics

The demo part of the meeting should be open to all. Any stakeholder who wishes to participate should be able to attend. Borrowing from a technique I developed with Greg Oehman at Salesforce, we always

---

29  https://rework.withgoogle.com/blog/five-keys-to-a-successful-google-team/

record the demos and post them somewhere afterwards (wiki, Google Drive, etc.) so that anyone who wasn't able to attend will be able to see the demos. This is critical if you have a globally distributed organization where time zones make attendance for all a challenge. However, the feedback from those folks can be invaluable to make sure we deliver the right work on time. Again, we're trying to create an environment that maximizes transparency.

In the retrospective part of the meeting, only the team members should participate (in Agile terms, only the pigs). There should be no executives or project managers attending this part of the meeting. It's strictly for those who need a psychologically safe space to have open and honest conversation in order to move the team forward, or to discuss problems without any fear of retribution or interference. Team members fill out a shared document or perhaps their own document with their thoughts about the iteration (what went well, what didn't go well, what could be better). Then each member in turn has an opportunity to read their contribution and explain in greater detail so that they know that they've been heard. During this section of the meeting, clarifying questions can be asked, arch meeting agenda items can be added, etc.

When holding the demo/retro at the end of a week, we like to have the team spend the rest of the day working on documentation, testing in staging, development, etc. Basically anything that doesn't touch production before a weekend.

## SRE Meeting Plan Conclusion

Finding a cadence at which to work as an engineer can be difficult. As engineers are generally averse to meetings, oftentimes we wind up with sporadic meetings and a lot of people who are unclear on their priorities and goals. On the other side, we can find ourselves in environments that are extremely meeting heavy, and engineers are often left wondering when there will be time to actually do the work they believe they were hired to do. The establishment of only necessary meetings, at specifically defined times, allows engineers to plan their time to minimize context switching, and to maximize the time invested in their meetings with one another.

This plan is certainly not a one-size-fits-all solution. It's deliberately broad and flexible so that you can modify it to fit into your organization, while it's prescriptive enough about the purpose of each interaction to allow for different implementations that accomplish the same goals, namely: transparency, collaboration, agility, and effectiveness.

I hope you're able to use it to advance the capabilities and success of *your* SRE teams.

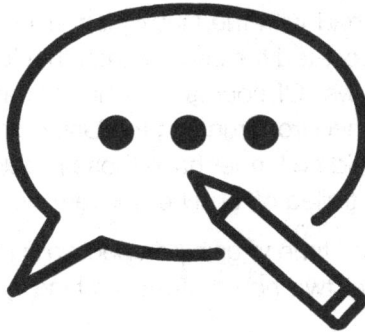

I'd been working with a portfolio company that had already broken its organization into two-pizza teams. After weeks of trying, the leaders were finally ready for their first architecture meeting. They decided to bring their most senior engineers together for the first few meetings and then expand to a wider audience after they'd achieved some momentum.

After some initial discussion about how the meeting should run, what the objectives were, and what the plan was going forward, the discussion started. One of the engineers asked about a system that had fallen into disrepair and disuse. "This used to be operational, why did we stop using it? If we were going to replace it with something that does similar things, what would be the pitfalls that we'd have to watch out for? What could we do to make it better?"

These types of discussions had not been happening in this organization. Despite the fact that the engineers were very talented, everyone had been working on their own projects and initiatives. The information started flowing, ideas were shared, decisions were debated, it was like watching magic happen right before my eyes. The dedicated time gave people the opportunity to step back and look at the larger picture from a different perspective.

It gave them the opportunity to look at the problem like George. George had been part of a company that had been acquired, and he explained to me something that he'd learned. He told me how Windows was a much better server platform than Linux. I was curious and asked why he thought so. He said they'd hired a new CTO from Microsoft who had them rebuild the entire server farm from Linux to Windows, so many of the problems they'd had with the Linux platform they no longer had. I asked him if they'd repeated the same things they'd had problems with on Linux onto Windows. Of course they hadn't, he replied. "We took the lessons we'd learned from running the Linux platform and changed things so that we would no longer have those problems on Linux…," at which point his voice trailed off and he got very silent and humble.

Setting aside dedicated time to gain perspective is like the old computer science joke: Never let two hours of researching get in the way of two months of coding.

# Quality Engineering Teams

DevOps is not just about Dev and Ops. As we'll see in the next chapter, we can apply DevOps principles to other parts of the organization in general (e.g., I've applied DevOps principles to a COVID vaccination flow), and Engineering in particular.

One area that often gets disrupted during the private equity buy-and-build strategy is quality engineering. Quality *engineering*, as distinct from quality *assurance*.

For a company like Salesforce, delivery with quality (and likewise trust) is an extremely important value for one of the world's largest software vendors. So much so that training in quality practices is (or was) a regular part of developing the company's engineers. One outcome of this emphasis was that I was able to give a talk at the very first DevOps Enterprise Summit called "Building Quality in at Scale" with

Reena Matthew,[30] with Reena Matthew, a principal architect in quality engineering at Salesforce.

One of the things that was stressed in the training was that we cannot *assure* quality in software. If you work in a toothpaste factory, you can take a statistically significant sample of the tubes of toothpaste that come off the production line. If enough tubes demonstrate that they're within the specifications identified, because they all came off the same line, we can assure the quality of that run. That is, we can mathematically prove that we have achieved a level of quality that matches specifications in that particular batch.

As much as we try to mimic a production line in software using our CI/CD pipelines, there is no way that we can assure the quality of the software produced. It would be like trying to do the same for toothpaste if the ingredients changed constantly. There are different developers, different needs for each method written, different languages and libraries, different ways to implement the same thing. In addition, we cannot randomly sample lines of code coming through the pipelines and make any determinations about the quality of the codebase overall. There is no way to mathematically prove anything. Adding 100 lines to the code could be catastrophic. Reducing a method from 150 lines to 100 could change the margin for a business by hundreds of millions of dollars.

**DevOps Patterns for Private Equity**

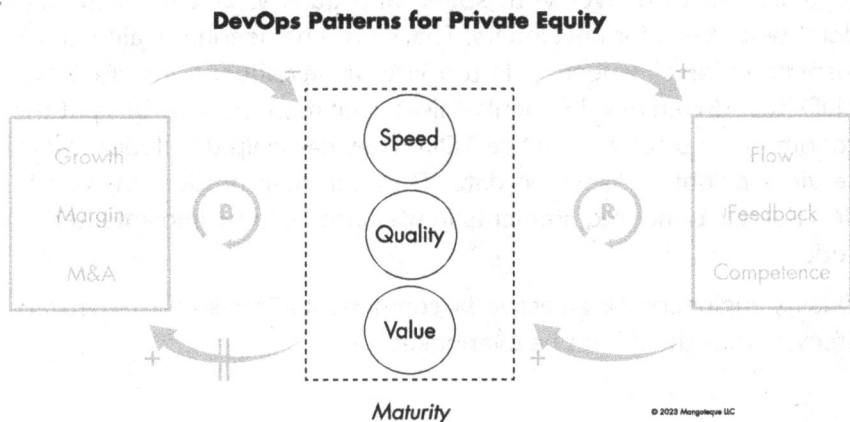

| Growth | | Speed | | Flow |
| Margin | B | Quality | R | Feedback |
| M&A | | Value | | Competence |

*Maturity*

© 2023 Mangoteque LLC

---

30  https://videos.itrevolution.com/watch/524582913/

In a DevOps context we want to deliver value with speed *and* quality. This is where quality engineering comes in. Because we cannot assess the quality of the code by sampling lines, many organizations still try to develop some type of assurance by involving human QA folks who review changes coming out of development. But the moment humans become involved, we start to lose our speed. That's not to say there's no role for exploratory testing in quality engineering, but that it cannot hold up the delivery of software. Automated software testing shortens and amplifies feedback loops (the second way). Human testing does the opposite. Like a flapping test, humans are fallible, they have bad days, they have fights with their partners....

Instead, quality engineers (QEs) should help to engineer systems that will enable code to be tested easily, comprehensively, and quickly. This is an essential component in delivering with speed and quality. Quality engineers can work with developers to establish testing frameworks for different languages, determine what should be tested at which stage of the pipeline, and create agreement on how these tests will be deployed, consumed, and reported upon. What makes sense as a unit test? What do we do about software the company has tucked in that was never written to be unit tested? What requires a virtual browser? These tests can include performance and security as well as pure functionality.

Quality engineering helps to put the conditions in place for the organization to deliver with speed and quality. Quality engineers don't write tests for developers. That's not engineering quality, that's inspection, and inspection is too late. If developers are practicing TDD (test-driven development), this is even more so. The QEs put the conditions in place to practice TDD. They can help developers mock services or obtain sanitized data. They can help product owners to describe customer requirements in ways that can be implemented as code.

Quality engineering is an essential component in the software delivery lifecycle that should not be overlooked.

# *Show Me The Tests*

When working with portfolio companies, the question of how and when to do testing comes up often. How do we handle code that wasn't written to be tested?[31] How do we know if a release is ready for production? How do we know if a product meets the specification?

If we want to deliver software with speed and quality, automated testing is essential.

## Release

I was sitting in a go/no go meeting for the release of a piece of software. That sentence should already tell you that, at the time, this company wasn't releasing often. Every release was a "big bang" affair. The different groups involved (Development, QA, Operations, etc.) were each given a chance to give a thumbs up or down to the release. When it was each leader's turn, they would explain how they felt about the release and then cast their vote. Like a jury.

I've coached leaders on how to ask their employees about what makes them frustrated and what challenges they enjoy. I've encouraged team members to be open and honest with each other, to talk about their feelings. I think feelings are very important.

---

31  https://podcasts.apple.com/us/podcast/devops-with-dave-mangot/id712327513?i=1000567411483

I think feelings are very important, unless we're talking about whether or not to release software. In that case, show me the tests. Did the tests pass? Great. Do the release. Did the tests not all pass? Then we need to go back to the software and fix it until they do. There's little room for feelings in software release. Either the tests pass, or they don't. If we don't have tests, then that's the problem to solve.

There can be unit tests, integration tests, security tests, performance tests. At Salesforce we had hundreds of thousands of tests that ran against the software. We had teams dedicated to keeping the farms of testing infrastructure up and running. The more testing you do, the more confidence you'll have in your software when it's released.

Part of the suite of tests to accept that the software is suitable for release are acceptance tests....

## Product Specifications

I've talked to, or worked with, a number of portfolio companies that are involved in financial calculations (taxes, payroll, etc.). One of their concerns has been that they might release software that does these types of calculations incorrectly.

These are perfect candidates for tests.

## Pre-Production

In the development of the software, we know what we're trying to calculate. If we were to practice test-driven development (TDD), we would write the tests *first* and write the software until the tests pass.

If I need to calculate a 5 percent tax rate on $100, the software had better calculate $5 as the tax. If it doesn't, then that software isn't ready for production. The tax, or payroll, or whatever is necessary in that feature, is part of the product specification. Part of the product specification can be the acceptance tests as discussed above. If the acceptance tests aren't accepted, then that software isn't ready for release.

The developers need to write these tests as part of the process of development. They cannot wait until someone from QA has a chance to write them or execute them. Part of being a developer is writing tests. As my friend Jonathan Hall says, "If you can't write code and verify that it's doing what you expect it to do ... you're not a programmer. You're a guesser."[32]

## Production

If an especially tricky calculation is required, we can "dark launch" the code into production as a way of testing. The code is released but not used for anything. The inputs and outputs of the calculation are written to some datastore that's not read by the primary application. This is to account for strange things that the user may input that we didn't take into account in our acceptance tests.

We can run this dark-launched code in production until we're satisfied and confident that the calculation performs as expected. Only then can we permit that code to be used for our customers.

# Security and Compliance Teams

One area of the software delivery lifecycle that's rarely overlooked is that of security and compliance. We'll dive much deeper into security in the next chapter, but since we're building high-performing internal product teams in operations, Ops teams have other non-development customers as well: security and compliance teams.

While security may have been an afterthought at one time or have contributed by saying "no" to everything, security is now recognized in the DevOps movement as an important component of delivery with speed and quality. The primary role of security is not to eliminate threats but to reduce risk and facilitate robust response to incidents. As such, there needs to be a partnership between our development and operations teams on the one hand and security on the other.

In order to reduce risk, often the security team will need agents to be deployed to different parts of the infrastructure to collect security

---

32  https://techhub.social/@jhall/109948986693864259

telemetry to be used for analysis. How should these agents be deployed? If we were to take a project approach, our teams would run some kind of orchestration tool that would install the agent on the available hosts. But what if a host isn't available? What if a host hasn't been deployed yet?

If we take a product approach, then the agent will be built into each relevant piece of infrastructure as part of how it's instantiated. It could be container images or machine images. It could be part of the configuration of network gear. The important part is that the agent is not deployed as part of an activity that has a beginning and an end. It's part of the lifecycle. What if someone were to forget to deploy the agent on new infrastructure? Should that infrastructure just not be secure? By taking a product mindset, we don't rely on humans (again, fallible!) to remember things. It's best to use humans to design how to incorporate the agent into the lifecycle of infrastructure, with its requisite improvements and configuration changes. The repetitive, boring work of ensuring that those activities are performed is left to the computers. This parallels our ideas of testing performed in Quality Engineering.

## Compliance

In order to ensure a robust response to incidents, Security can work with Ops to ensure automated governance, which we introduced earlier. The things auditors care about most are:

- Who did what?
- When did they do it?
- What allowed them to do it?
- Is everything configured as you say?

Before the days of infrastructure as code, audits like this would require documents describing how the infrastructure was configured, and then representative samples of machines (like toothpaste!) would be selected to demonstrate that the equipment was configured as described.

If the selected machines demonstrated what had been described, then the audit could proceed. If they were different, then remediation was necessary, and more investigation was warranted.

With the rise of infrastructure as code, this elaborate game was no longer necessary. I have often described configuration management tools as compliance engines. The tool describes the desired state of the machine, and that state is continually enforced upon the regularly scheduled or even imperatively initiated invocation of the tool. Anything out of specification is brought into alignment. If any of your engineers have noted the disappearance of a cloud resource after a Terraform run, this is a compliance engine in action! If we're using immutable infrastructure, the same concept applies. Because the infrastructure is immutable and ephemeral, we can destroy and replace instances to be sure everything is configured as described and cannot be changed.

All these activities are logged: additions, deletions, and changes. Because the logs can generate reports, we can demonstrate to the auditors the state of the infrastructure. We can extend the pattern to include deployments, code check-ins, and CI/CD tests. If the logs are immutable and/or cryptographically signed, we can demonstrate to the auditors who did what when.

When the security team works with the Ops teams to integrate these patterns into the different elements of the software and infrastructure delivery pipelines, they know about changes when they happen. They're able to look for unapproved changes (i.e., possible intrusions) and are able to give the auditors what they need. Because this is all automated, it's automated governance.

# Cross-Functional Teams

We've talked about SRE teams, quality engineering teams, and security teams, but what about cross-functional teams? These are teams that have representation from multiple disciplines so they can execute on solutions without waiting, handoffs, or interruptions (leaving self-service aside for the moment).

In short, I haven't seen them in portcos. It would be amazing to have an SRE on every development team working on resilience patterns and incident response. In the Googles, Amazons, or Salesforces of the world, with tens of thousands of employees, these teams exist. At a certain scale, the centralization we create with two-pizza Ops teams becomes decentralized once again. I simply have not seen them with my clients. Even if they were to fund having fully cross-functional teams for every product, there need to be methods of communication to make sure that work isn't duplicated all over the organization. The biggest waste of resources would be to have each SRE develop its own local way of deploying to Kubernetes only to find that we now have eight different ways. This means that there need to be channels of communication between these functional areas that ensure that each person in that role throughout the company is aware of these initiatives (communities of practice). It's not that it can't be done, it's that it's expensive and requires a lot of intention.

Instead, what I see much more often is an emphasis on DevOps collaboration. Teams meet regularly to understand needs and objectives and then empower others to be able to deliver without obstacles *as if they had those team members on the team*, also known as shifting left. In *Team Topologies* parlance, the stream-aligned teams work together with the platform teams, complicated subsystem teams, and enabling teams so that the stream-aligned teams can continue to deliver to the value stream with ease. Almost as if they were a virtual, cross-functional team.

This may limit some professional development since engineers have less in-depth exposure to other disciplines. But it also encourages more depth (à la T-shaped engineers) within a team, which allows for high-performance execution across the organization.

## *Attrition*

The chaotic nature of 47 lone wolves, each working on their own projects, causes a considerable amount of stress throughout the organization, especially in Ops. People will put up with a certain amount of stress during big transitions, but they won't tolerate it indefinitely, which leads to attrition.

In their own words:

- "With the stress, it's not worth it"
- "We're in a panic mode all the time"

Many engineers decide that they would be better off leaving the organization so they could have less stress and make more money. In some organizations, I've seen the majority of engineers leave. One engineering manager expressed concern that people were leaving "faster than they can be replaced." This is not revenue protecting!

When work is organized by project instead of product, and an engineer leaves, their knowledge and expertise leave with them. When engineers work on a team, it's the team that's responsible for that knowledge and expertise. When someone is an expert, they can pair with or mentor other team members so the knowledge is spread throughout the team. If there's only one person who knows about database system X, then every time system X has a problem, there's only one person to call, day or night, on vacation or off, to deal with the problem.

We can retain our engineering talent if the job of the team is to make the hard things easy so that they can work on bigger and more challenging problems (read: improving numbers both above and below the line). This is using the engineers to their full potential to help the business meet its objectives.

I've seen organizations where massive cost savings and development efforts were achieved when simple things like deploying servers and applications became trivial, so that engineers could work on more difficult problems. Employees who are challenged in their jobs want to stay, even if they have to take a little less money. Getting paid slightly more to be bored, or worse, stressed, is not worth it.

## *Onboarding Costs*

Teams that work to deliver products write documentation. They figure out ways to cross-train members. They develop runbooks for when there are outages. These teams use the occasion of bringing on new engineers as an opportunity to look for gaps in the materials they've prepared, to seek corrections to things that may be outdated, and to develop new materials. Teams that treat each work item as a project try to finish those work items so they can work on the next project.

The amount of time it takes to onboard a new engineer into an organization can have significant impacts on engineering organization budgets. I worked with one engineering leader who told me that they felt that anyone on their teams making less than $170,000/yr was considered "at risk" of leaving to obtain a better salary. He explained that the typical onboarding time for a new engineer was one year!

Considering a minimum salary of $170,000, plus the commission that must be paid to a recruiter, the time to interview, and the work involved to negotiate and hire, the cost to bring on board a new engineer can easily exceed $250,000. Senior engineers should be able to onboard in two months, and more junior ones should be able to accomplish the same in three. That means that this organization was wasting 75 percent of their time and money for each new engineer. If they were to hire five engineers in a year, that's almost a million dollars of waste, plus time.

Working with a project instead of a product mindset not only causes good engineers to leave, it's damaging on arrival as well.

## *Financial Performance*

Many of the topics we've discussed in this chapter are directly related to operating expenses, retention, and delivery. There are many actions that can be taken to improve financial performance.

Self-service tools for engineering allow product teams to deliver with increased velocity. Nobody wants to wait for their tickets to be satisfied by the Ops team. As a development manager told me: "I have lost developers because of this. They are frustrated." Increased product velocity allows teams to deliver more of what the customer wants in the same amount of time, which helps to drive revenue, market share, and customer retention.

Improved efficiency and optimizations by the Ops teams can improve margins. Instead of being a cost center, these teams are partners in driving innovation! When the team is building infrastructure as products, they can automatically tag their cloud resources to make it easier to determine which products are spending how much money, and then help those teams to capacity plan and optimize. They can go back to the cloud providers and negotiate for discounts. None of which is possible without a solid understanding of their cloud spend.

When monitoring and security agents are deployed uniformly, the team has the ability to detect and recover from incidents more quickly. Less and shorter downtime will improve the customer experience and increase retention. I talked to a portfolio company that was in danger of losing their second largest customer. They were having a lot of downtime, and because of their lack of monitoring, it was hard to prove whether the problem was on their end or the customer's side. We began to look at how the infrastructure was deployed, and because the customer was so large, they had made many one-off exceptions to the way things were run in order to satisfy customer requests. They were essentially running their product completely differently for this one customer, and instead of being appreciative, the customer was furious that the portco was delivering such poor availability. Sometimes I wonder whether portcos hire me so the CEO can get out of the business of apologizing to customers.

When teams deliver products using infrastructure as code it forces consistency. Consistency leads to supportability, supportability leads to availability, and availability leads to happy customers.

Lastly, when the development and operations teams are delivering at a high level, performance increases. Traffic patterns are well understood, and measures can be taken to make sure products are delivered within performance specifications. There are legendary Silicon Valley stories about the impacts of performance degradation.

According to Yoav Einsv of Gigaspaces, "Amazon found that every 100 ms of latency cost them 1% in sales. Google found an extra .5 seconds in search page generation time dropped traffic by 20%.[33] A broker could

---

33  http://glinden.blogspot.com/2006/11/marissa-mayer-at-web-20.html

lose $4 million in revenue per millisecond[34] if their electronic trading platform is 5 milliseconds behind the competition."[35]

Highest performers are twice as likely to meet or exceed organizational performance goals, including financial performance.

34 https://research.tabbgroup.com/report/v06-007-value-millisecond-finding-optimal-speed-trading-infrastructure
35 https://www.gigaspaces.com/blog/amazon-found-every-100ms-of-latency-cost-them-1-in-sales

# There's No Such Thing as DevSecOps

*"'Shift left' is the purported mantra of DevSecOps, but what it means too often in practice is shifting obstructionism earlier into the development process."*

— Kelly Shortridge, Fastly

If I were to say the name Kleenex, you'd probably think about tissues.

If I were to say the name Coca-Cola, you'd probably think about soda.

If I were to say the name DevOps, you might think about a lot of different things by now.

We know that the DevOps movement is about optimizing the flow of work through the system, getting fast feedback, and practicing and experimenting in order to improve outcomes. As we often say, building with speed and quality.

**DevOps Patterns for Private Equity**

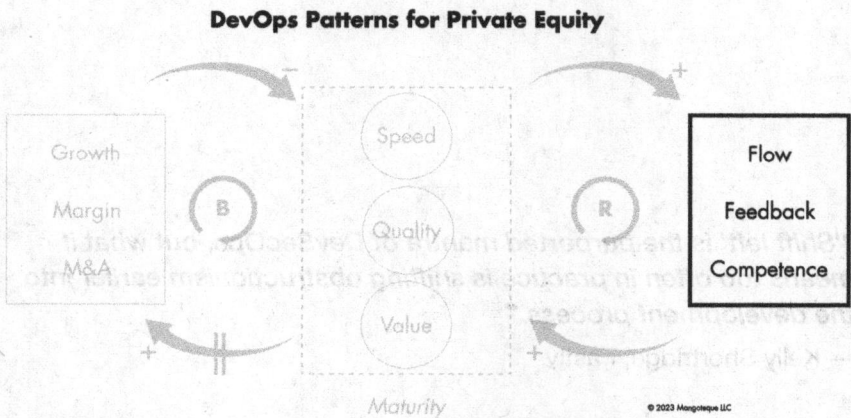

If security doesn't do its job, and the company is hacked, we haven't improved our outcomes. In fact, we've made them worse. If marketing doesn't do its job, and no one knows about the product, we haven't improved our outcomes.

One curious side effect of the term "DevOps" is that people often get hung up on the term and try to highlight other areas of the business that are also important. A common example is the term DevSecOps, which has also been called Rugged DevOps. Saying we need to focus on DevSecOps is like referring to the department of redundancy department. DevOps is no longer about just Dev and Ops. It's a brand name at this point for an international movement to bring humanism and technology together to achieve better business outcomes.

Does that include marketing? Yes. Security? Yes. Legal? You betcha.

Quality engineering? Of course.

DevMarSecLegQEOps?

In this chapter, we examine DevOps without focusing on Dev or Ops. DevOps principles can be applied to many parts of the organization to build better businesses and achieve better outcomes – in this case, with security built in.

# Quality

I was lucky enough during my time at Salesforce to work with the "Father of SSL," Taher Elgamal. While at Netscape in the 1990s, Taher developed the ability to encrypt web communication. None of us would feel comfortable sending our credit card information, or bank logins, across the Internet without the groundbreaking work of Taher.

I loved listening to Taher speak about anything security related. If there was an opportunity to hear him offer his opinions on a security topic, I always tried to find a seat in the room.

During the time Taher and I overlapped at Salesforce, the DevOps movement was really taking off, and I was the one leading (with a *lot* of help) the adoption of DevOps principles. This was also a time of massive hiring within the organization, and we were flying roomfuls of new engineers to San Francisco for orientation multiple times a month. For each of those orientation sessions, I would give an "Introduction to DevOps at Salesforce" talk to the incoming class.

One day, Taher came to give an orientation talk to the new hires. I immediately found a seat in the back of the room. During the talk, Taher said something that changed the way I think about security forever.

*"Security is an aspect of quality."*
– Taher Elgamal

Now, on its own that might not seem especially earth shattering. But I've also studied someone who is often called the "Father of Quality," W. Edwards Deming. As we read in an earlier chapter when talking about change advisory boards:

*"Inspection does not improve the quality, nor guarantee quality. Inspection is too late. The quality, good or bad, is already in the product. As Harold F. Dodge said, 'You can not inspect quality into a product.'"*
— W. Edwards Deming, Out of the Crisis

Well, if inspection is too late, and the quality is already in the product, and security is an aspect of quality, then, we can't inspect security into the product either! All of the penetration tests, and blue, red, and purple team hacking, all the fancy presentations at DefCon are useful, but they do not make something secure. Only by building security into a product can we reduce risk (i.e., become more secure) and improve quality.

# The Three Ways of DevOps and Security

Because we know that the goal of DevOps is not to trade off speed and quality for one another but to have both, the role of security is to enable value creation at speed with high quality (i.e., high security). Through this lens, security is no longer the party of "no" but instead active, engaged participants in the activity of value creation.

Before Gene Kim described the three ways of DevOps in *The Phoenix Project*, he described them in a blog post.[36] I've always thought about them as one of the best ways to help people understand DevOps. Let's use the three ways to describe how to build security in using DevOps.

---

36  https://itrevolution.com/articles/the-three-ways-principles-underpinning-devops/

**DevOps Patterns for Private Equity**

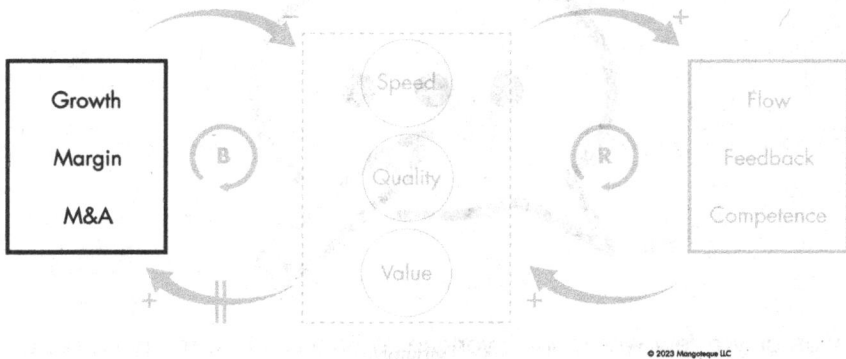

© 2023 Mangoteque LLC

# The First Way

*"The First Way emphasizes the performance of the entire system, as opposed to the performance of a specific silo of work or department ... always seeking to increase flow, and always seeking to achieve profound understanding of the system (as per Deming)."*

— Gene Kim

The first way is about applying systems thinking to the creation of value in our organizations. We are trying to avoid local optimization (i.e., local minima) in favor of global optimization (i.e., global maxima). This is straight from Eli Goldratt's *The Goal*, the inspiration for *The Phoenix Project*. One way to maximize our value creation is to shift things to earlier in the process.

I was only a few weeks into working at Salesforce in my new role as Architect, Infrastructure Engineering. I attended a meeting with a number of security folks, and the person leading the meeting was on a tirade about some security problem that had been deployed to production. They were ranting about punishing developers who wrote code with security problems, how they had to learn, how dumb they were for making the wrong choices.

Never one to be afraid to speak up, I posed a question. "If the developers were making bad choices, then weren't they put in a position for bad choices to be made?" If we offered them the right choices, and those choices were easier to implement than the bad choices, then no one would choose to make their life harder.

Make the right way the easy way and that's what people will choose.

Very few people in the room knew who this brash new architect was, and I received a lot of blank stares, and a number of people introduced themselves afterwards. Nowadays, it's not even a question that Salesforce's development pipelines are built with those principles in mind.

We want to build security in as a foundational element in the way software is developed. We don't want to apply patches to our production instances after the fact. That's expensive. We want to build the software and machine images in a secure manner. Let's look at some examples:

- *Pre-approved libraries*: By now many people have experienced the log4j or OpenSSL patching mess. Using pre-approved libraries doesn't necessarily help against future problems, but it will help to ensure that old problems aren't introduced to new software. The provenance of each piece of software released should be tightly controlled. The entire software supply chain should be considered with chain-of-custody principles. When my teams used to build Amazon Machine Images (AMIs), they would mirror the upstream Linux provider's repositories into our own private copy at specific intervals. This way we knew exactly what was being deployed onto our instances instead of continually mirroring and chasing a moving target. All the building blocks of the software and infrastructure should be well-known and blessed.

- *Blessed artifact repositories:* Because the provenance should be tightly controlled, there should be blessed artifact repositories from which people build their code, AMIs, or container images. It's estimated that between 63 percent and 96 percent of third-party container images contain known vulnerabilities.[37] Do not let engineers base their applications on random images pulled off the Internet. In order to be more secure, provide images for people to use so that the foundations of your infrastructure are secure and easy to consume. Make the right way the easy way.

- *Software bill of materials (SBOM):* One way to keep track of the provenance of software, and to quickly determine what needs to be patched in the case of a vulnerability disclosure, is through a software bill of materials. At the time of this writing, SBOMs are in their infancy in the industry. Their use isn't widespread, and the tools are rudimentary. With new rules by the U.S. federal government and the recent spate of software supply chain attacks, the importance of SBOMs will only increase. It's important that all SBOMs be computer-readable. The complexity of chains of SBOMs with software built on top of software quickly outpaces the ability for humans to consume, track, or reason about. If this is not yet a practice in your organization, now is the time to prepare.

---

37  https://www.zdnet.com/article/96-of-third-party-container-applications-deployed-in-cloud-infrastructure-contain-known-vulnerabilities-unit-42/

- *Tabletop exercises*: Before any new major additions are made to software or infrastructure, it's important to discuss elements of how things will be built. Where will secrets be stored? How often will they be rotated? By what mechanism? How often will things be patched? By what mechanism? What will we do if a vulnerability is discovered? The answers to these and other questions should be discussed before anything is built. If your organization has standard ways to deal with these things, fantastic! If not, now is an excellent time to develop standard practices. Fewer risks will make it to production, and remediation will be faster if everyone understands how these things are handled. If you've done a number of acquisitions, and each team has its own way of doing things, there will be a price to pay — complexity is not free.

There are many more opportunities to improve the secure foundations of our software and infrastructure. Ultimately, we don't want to teach developers to do the job of security, we want to enable developers to build secure software without having to be security experts. In other words, shift left.

## Shifting Left

Shift left! Shift left! they say. This is a common refrain when people are talking about DevOps. Deming taught us that we have to build quality in, we can't inspect quality in at the end, so we need to shift left.

So teams set off to shift left, not sure exactly what this means. Often it's approached as though it will mean that more of their work will be given to the developers. They will finally understand what we go through! They will finally learn about our jobs!

- Operations folks believe it means that the developers will become experts in infrastructure and will be on-call for every outage.

- Security folks think it means that developers will have to keep in mind every XSS, CSRF, SQLi and any other vulnerability possible while still writing high-quality code.

Shifting left doesn't mean that other people do our jobs. We still have to do our jobs, and the developers still have to deliver business logic for the customers.

*Shifting left means making it easy*
*and safe to do things earlier in the process.*

# The Platform

In the early days of infrastructure as code we spent a lot of time trying to figure out who would be writing infrastructure code. The Ops folks didn't have a lot of experience writing code, and the developers probably needed to understand the infrastructure. So we would demand that developers needed to learn *our* tools, and *our* jobs, so that we could deploy their code and infrastructure.

Except that's not shifting left. That's making the developers do our work. With the rise of platform teams, we can see that we need to provide a way for developers to deploy on infrastructure without being concerned with all the details (e.g., IAM, security groups, routing), while still having the ability to do it themselves (*easy*) and without much opportunity to break things (*safe*). That doesn't mean they need to learn CloudFormation.

I once worked with a very high-performing software development team that had deployed their own infrastructure before they'd hired any Ops folks. They spent the bare minimum of time on the infra because their job was *shipping software*, not worrying about network interfaces. When we shifted left, they could easily and safely deploy the necessary infrastructure to build their services, and they didn't need to write infrastructure code!

## We Don't Want to Do It Either

I was working with a client whose Ops team felt it was toil to create the user accounts in Gitlab for all the different development teams and manage all the group assignments, new hires, and departures. They decided that they would shift left and give permission (and detailed instructions describing all the steps required) to the development managers. Empowerment!

The response from the development managers was fairly predictable:

1.  We can see why you don't want to have to do this.

2.  We don't want to do your job for you, we already have our own concerns.

The Ops team was simultaneously furious and disheartened. They thought the managers would be grateful. They only realized after the fact that this wasn't eliminating toil, this was distributing toil to the edge.

To understand how to truly shift left, we discussed when user accounts were created and where. It turns out that they had an Active Directory setup, and GitLab even had an LDAP integration![38] Sure, there would be configuration needed and probably some glue code had to be written to get everything set up properly, but now user management for those users would be essentially automatic. As we say in Agile: maximize the work not done.

## Really Shifting Left

In order to really shift left, we need to enable people to achieve their objectives safely and early, without needing to have our expertise in addition to the expertise they were hired for. It's not just about moving the work earlier in the process.

---

38  https://docs.gitlab.com/ee/administration/auth/ldap/

There are lots of different groups that can empower developers to work quickly and safely:

- Infrastructure
- Security
- Quality Engineering
- Marketing

How you help the whole organization safely and easily ship higher-quality software is up to you.

## The Second Way

*"The Second Way is about creating the right-to-left feedback loops. The goal of almost any process improvement initiative is to shorten and amplify feedback loops so necessary corrections can be continually made."*
— Gene Kim

Building software on a secure foundation means that we can have more trust in the things we build. But humans are fallible, and no system is perfect. A foundational element of DevOps, Agile, and Lean is feedback. If we're able to incorporate feedback quickly and easily, then "necessary corrections can be continually made." An organization that incorporates feedback quickly and easily becomes a learning organization. Learning at speed (with quality) is a tremendous business differentiator. In order to build quality (i.e., security) in, we need to be able to use our feedback to make and keep our products more secure.

We've already discussed ways to improve our foundation. Now we turn our attention to some ways to "shorten and amplify feedback loops" to make our organizations more secure.

Catching bugs and security problems early in the process is faster and cheaper than catching them later on. If you're a developer and you write some code and learn 10 or 15 minutes later that you've written a bug (roughly the amount of time Humble and Farley described in their book *Continuous Delivery*), it's easy to fix. You just wrote the code, you'll remember the choices you made and why. If you write some code and you discover the problem three weeks later when it's deployed in production, it's difficult and expensive to fix. You'll need to re-familiarize yourself with what you'd done. You'll need to run the entire gauntlet of testing before production. If it's critical, you may need to get executive approval to push the fix outside of normal procedures.

Here are some ways to help you shorten and amplify security feedback loops:

- *Security code scanners:* Years ago I watched a presentation by Square that I think exemplifies this concept. Developers would check code into revision control, and various security scanners (e.g., static code analyses) were run on the code to look for security problems. If any were found, developers would get an email explaining the problem with an example of how other developers had fixed similar problems in the past. It was a quick feedback loop so that necessary corrections could be made. Fast forward a few years and these same types of tools are even available in the IDE, before code commit! The tools will suggest alternative commits that are more secure than what had been written. That is a short feedback loop!

- *Continuous hacking:* Some banks have taken this concept even further.[39] Upon code commit, containers are built with the code just committed to the CI/CD pipeline. The containers are deployed into a representative test environment and then various security tools attempt to "hack" the containers looking for security vulnerabilities. If vulnerabilities are found during this "continuous hacking," the code is not allowed to proceed to the next step in the pipeline, and the developer is given an opportunity to fix their vulnerabilities.

- *Internal libraries:* We've found a number of places where our developers make the same security mistakes, or we've identified some areas used in the codebase that are "security critical," like where we're retrieving information from the database or storing

---

39  https://fluidattacks.com/services/continuous-hacking/

credentials. Shortening the feedback loop is great, but how do we amplify it? Just as we discussed in the previous chapter about productizing our services, we can productize our learning through internal libraries. Just as no sane developer would roll their own cryptographic algorithms, it doesn't make sense for developers to continually roll their own database access. In our internal libraries we can create methods in partnership with security for things like database access. Developers can simply call a method to insert data in the database that has been sanitized and checked for SQL injection vulnerabilities and has the access credentials handled in a secure manner, etc. These types of internal libraries amplify the work we've done to secure the environment by making the right way the easy way and encouraging its adoption throughout the codebase or even across multiple products.

- *Monitoring*: Once our code is deployed and running in production, the entire environment needs to be monitored to ensure that things are working as expected. When we say "as expected," we mean that there is no unexpected network traffic — that we are blocking malicious code either at the WAF (web application firewall) or in methods dedicated to this purpose — but also that access patterns to critical resources are as expected. For instance, if we have a bucket on AWS S3 that holds files that are only supposed to be accessible from internal networks, we can continuously test and probe to see if we can access those files externally. People make mistakes, policies are applied to resources inadvertently.

If we can continuously monitor not just our application performance but also security through logs, probes, and active testing, then we can both prevent many issues and be better prepared when breaches occur.

## The Third Way

*"The Third Way is about creating a culture that fosters two things: continual experimentation, taking risks and learning from failure; and understanding that repetition and practice is the prerequisite to mastery."*

— Gene Kim

We've invested time and energy in securing our build pipelines. We've implemented SBOMs, we're continuously monitoring our infrastructure through agents our security team has deployed in collaboration with Infrastructure Engineering, and still, security incidents will happen.

There is a well-known (absolutely hilarious) talk from the 2014 Monitorama Portland conference, where James Mickens, then of Microsoft Research, discussed threat modeling in security.[40] In his view, there are two possible scenarios: Mossad, and Not Mossad. If you're being attacked by Not Mossad, you have a chance of repelling the intrusion. If you're being attacked by Mossad, they have commandos and helicopters and fancy guns that can shoot around corners so you might as well give up.

Of course, while Mickens was playing to the crowd for laughs, the fact remains that keeping a well-resourced and determined attacker from gaining access to your data for ransomware or other purposes is difficult, and at some point it's very possible there will be a breach.

If we've invested heavily in our foundational elements, gaining access will be more difficult. However, no amount of container patching can defend against a well-executed social engineering attack. If we've invested heavily in our monitoring capabilities, hopefully we'll know when there's an incident.

Noted security researcher Kelly Shortridge has suggested building what she calls a "bamboozle layer" into container images. This filesystem layer will have some files with attractive names like *secret_passwords*.

---

40  https://vimeo.com/95066828

*txt* or *executive_compensation.csv*. If any files are accessed on that layer, she recommends crashing the container immediately and declaring an incident. There is no legitimate reason for anyone to access files at that layer.

*"An incident is something you've already invested, what you need to do is get a return on that investment."*[41]
— Sidney Dekker, human factors expert

Once we've had an incident, we'd like to remediate it, and then shorten and amplify our feedback loops to make sure the next time we have an incident, our response is even better.

Learning from incidents is great, but they're costly both in terms of time and resources. However, they're still a great source of learning. What is an organization to do? Have more incidents?

The third way of DevOps tells us to foster two things: *"continual experimentation, taking risks and learning from failure; and understanding that repetition and practice is the prerequisite to mastery."* Of course, telling security folks to take more risks is sure to earn you side-eyed glances, but there are ways to do so safely. When Jesse Robbins was at Amazon, he used to call himself the "master of disaster" for running simulated failure exercises called Gamedays. Google runs similar exercises called DiRT (Disaster Recovery Testing). We can apply these same types of exercises to simulated security incidents (and simulated non-security issues as well!) with the knowledge that repetition and practice is the prerequisite to mastery.

Here are some Gameday exercises:

- *Leaked credentials*: You, the CTO, walk in on a Monday morning to find out that the credentials for your production databases have been posted on social media. What should you do? Should you even have credentials that are not rotated multiple times a day? What's the proper response? Yell at people? If we practice for these

---

41  https://safetydifferently.com/just-culture-the-movie/

scenarios, engineers (who are closest to the information) have the opportunity to propose and try out solutions that are superior when they're not firefighting. When it comes time for the actual firefight, it will either be a non-issue or the team will be better prepared.

- *XSS:* You've hired a penetration testing team to attack the production website, and they come back with a long list of cross-site scripting (XSS) vulnerabilities. The development team is given a number of weeks to fix these and eventually reports back that they have all been remediated. You bring back the penetration team a few weeks later and they find just as many XSS vulnerabilities as before. They report that they merely changed a few bytes in the payload and the site is vulnerable again. A decision is made to develop some internal libraries that handle these scenarios properly so that the development teams don't need to play an endless game of whack-a-mole. Yes, this is a true story.

- *Emergency patching:* Heartbleed, log4j — we've mentioned them previously. At some point, major security vulnerabilities will be released for common pieces of software. It's not a matter of if, but when. How long does it take the organization to deploy updated versions of the software? How do the notifications occur? What resources are necessary for the remediation? The organization can either learn to perform these actions under fire, or they can practice them ahead of time, leading to mastery, so that when faced with these inevitable situations, they're prepared to respond in a calm and professional manner.

I regularly teach Gameday exercises to organizations that are launching production services, whether new services or products that are moving from the data center to the cloud. We develop a strategy for failure scenarios to be tested. We describe the failure (e.g., lost connection to the database), how we'll simulate the failure, the expected behavior, and the observed behavior. We repeat this exercise until the developers, the Ops team, and even security folks agree that the expected behavior and observed behavior are in alignment and documented. From this, runbooks, response procedures, alarms, alerting, etc. can be developed. Repetition and practice lead to mastery.

We know from safety sciences that a key to faster remediation of incidents is shared mental models. That is, the more people who have a good working mental representation of how the systems work, the better they can respond to unexpected situations. There's no need to wake Jessica at 3 a.m. for a cache failure if the entire team understands the functioning of said cache. If multiple people with shared mental models are working the incident, they're able to make better collective hypotheses about the causes (and remediation) of the problem.

When teams have applied the third way of DevOps, they have the ability to improve not only their mental models, but even the information they need to be effective. The three ways are broadly applicable outside of security to our entire SDLC.

As the engineer responsible, which would you rather receive?

> **ALERT - Unauthorized access from external source detected for S3 bucket.**

> or

> **2022-04-24 18:35 UTC. Security Alert - Unauthorized access from Geo DPRK detected on bucket cred-prod-access-ourcompany in region eu-ger-2b.**

Repetition and practice (and shortening and amplifying feedback loops!) is the prerequisite to mastery.

# Buy-and-Build

*"This venture into the improvement of quality and productivity is not a program with a beginning and an end, but a philosophy that directs efforts at all levels of responsibility towards the more effective use of the resources available to meet the needs of customers."*

— W. Edwards Deming, *Out of the Crisis*

The platform company was extremely active in M&A. In the past eight years they had acquired 20 companies. That was part of the problem. In the past eight years, they had acquired 20 companies!

The board was confident in its buy-and-build strategy, but engineering spent all its time just keeping up with all the acquisitions. Almost no effort had been made to actually integrate these companies into the existing platform, and sometimes they were in the same condition as when they were acquired, sometimes they had small improvements, and sometimes they were actually worse through attrition.

They had recently formed a data management team with a few database administrators (DBAs) to handle the technology sprawl. And sprawl there was. The team was responsible for multiple products running on the following systems of record:

- Postgres
- SQL Server
- MongoDB
- DB2
- Oracle
- MySQL
- Snowflake

One team needed to support 30+ development teams, 200 developers, seven database engines, and the rest of the operators.

The other operators, in addition to cloud deployments, were supporting seven data centers and five different monitoring systems. Some of the monitoring systems were only deployed in production environments. There was a data center closure project that had failed to achieve its objectives in three years, and then it was extended another three years.

To say that the integration of the acquisitions was challenging is kind.

**DevOps Patterns for Private Equity**

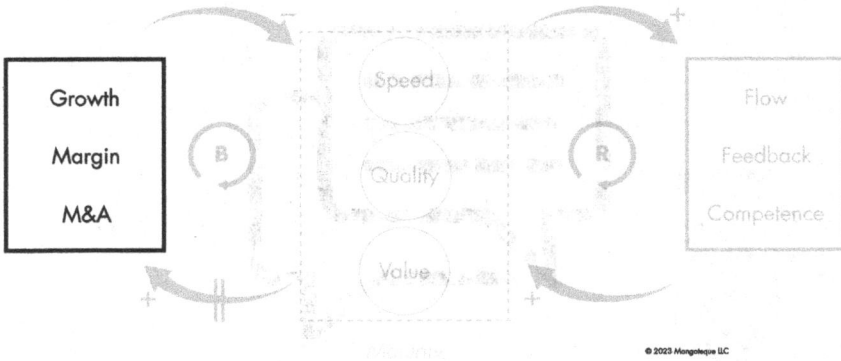

© 2023 Mangoteque LLC

# Impact of M&A

But why? Why is integrating acquisitions difficult? What makes it so challenging? What choices should we make? Where do things go off the rails?

I like to say that if you've ever worked with engineers, you know that almost every one of them thinks that the way they have implemented something is the best way. Combine a few companies, and all you have to do is convince each of them that other implementations might be superior.

As with most engineering leadership problems, the issues are twofold: technical and cultural. As with most engineering leadership problems, the easy one is the technical. Let's tackle it first.

## Technical Obstacles

# *Technical Debt is Not a Technical Problem*

Ward Cunningham developed the term "technical debt" to explain to some non-technical people why resources needed to be allocated for some engineering work. It seems like it would be a good metaphor, since many of us understand that debt often carries interest, and as that interest accrues, we accrue more debt. However, for business and finance people in the context of the business (as opposed to their personal lives), debt can be a wonderful thing. It can provide us leverage, which, among other things, allows us to have a longer runway for our startup. It can allow us to open offices in new geographies or expand our offerings onto new continents. Debt can be very useful. In certain business contexts, if we have a lot of debt we can package it up and sell it! Of course, this is not how tech debt works. It cannot be packaged up, sold, outsourced, or purchased from a SaaS provider.

Tech debt is also not inherently good or bad. Obsessively trying to eliminate all technical debt will almost certainly run quickly into the Pareto Principle, where 80 percent of the debt is a result of 20 percent of the problem, and trying to eliminate the last 20 percent of the debt is almost certainly not an economically prudent course of action.

At a startup this can be fatal. Startups are trying to achieve product market fit and break through to profitability. Spending a lot of time on hyper-clean technical implementation, rather than on survival, in this situation is downright foolish. Startups should take on *lots* of technical debt in the service of survival. After the product, the market, etc. are well understood, we can begin to determine the "right" way to architect, build, and continuously deploy the application.

In a business that has moved past the startup stage, tech debt can be a drag on the organization's ability to execute. In this context, tech debt is a business problem. This is the context about which Cunningham (one of the authors of the Agile manifesto) was concerned. If the business chooses to accept the reduced velocity in this case, then that is a *business decision*. It's a decision about the current conditions and the business's ability to compete.

I generally break down tech debt into two classifications: known and unknown. The tech debt that is a problem is generally the debt that is unknown. This debt comes from insufficient testing, documentation, training, planning, etc. This is often the debt that most often causes outages, weighs on development velocity, and causes us to miss our targets, creating further damage down the road. Some call this "dark debt."[42]

Known tech debt is a lot more like actual debt. If we know that tech debt is causing the problems listed, but choose to do nothing about it, that's a business decision. I hear so many engineers complain that management doesn't give them time to pay down technical debt and that they're suffering because of it. However, it's also the business that's suffering because of it.

When we choose to run 12 different datastores that can probably be merged down to a handful, that is a choice. In this case, we are knowingly making a decision to pay more, and to get less from our data storage layer.

----

42  https://medium.com/@allspaw/dark-debt-a508adb848dc

I firmly believe it's actually OK to take on tech debt as long as it's identified and acknowledged. We can acknowledge that we'll suffer reduced velocity in the short term while we make a business decision to do other things.

It's also important to note that this debt won't fix itself. We cannot expect engineers to work on it as a side project, or on their own time. Just as with monetary debt, there must be a conscious decision to pay it down, even if it's a little bit at a time every month. Just as with monetary debt, it will continue to accrue until it's unserviceable. Business decisions need to be made to maintain a velocity that won't damage the business itself.

When engineers know that they're being heard, and that their message is actually understood, they're much more willing to live with some amount of technical debt for a period of time. Accumulating debt (financial or technical) is best done with eyes open and with a smart payment plan that satisfies our financial, technical, and organizational needs and enables the business to flow.

I'd probably left the Scrum team more than a year before. I still knew the Scrum Master and had been on the team at its formation. I knew what technical debt they'd taken on at the beginning and had kept tabs on what they had accepted additionally over time.

"How's it been going?" I asked the Scrum Master.

"The team felt that there was too much tech debt and it was really getting in the way of their delivery, so we all sat down to do something about it." she replied.

"And leadership was OK with a few sprints being allocated to paying down tech debt?"

"Dave, this is an Agile team. We didn't go to leadership to ask for permission. As a team we agreed that the second Wednesday of every two-week sprint will be Tech Debt Wednesday for the foreseeable future. We're pretty confident that we'll be able to deliver on our commitments, and as the debt is paid down, deliver even more. Periodically we'll assess if this is working for us, and adjust as necessary."

The team went on to perform Tech Debt Wednesday for at least six months. Over time, the engineers were happier, leadership was happier, and their customers were happier that things that had been clunky and awkward to deal with in their interactions with the team had been paid down, improving everyone's velocity.

## IT Organizations

Having sprawl in functional areas is a challenge and makes them difficult to execute. Having duplication in organizations is downright wasteful. This sprawl isn't about who wants to write in .NET vs. Python. In IT organizations after M&A, it can mean having multiple departments that serve the same function.

IT (as opposed to production web engineering) is the way in which a company communicates. It's where work is prioritized in the ticketing system. It's where email is sent and delivered. It's how employees access internal systems.

Duplication in the IT systems is like running two (or more) separate companies. I worked with one client who two years after acquisition still hadn't merged the internal networks or the email servers. The fact that the company had so much trouble executing was due in large part to the fact that, despite being one company at the leadership level, it was two companies at the IT level. Integration of major IT systems should take months, not years, and should be a core part of the plan during acquisitions, not just to facilitate communication, but to make the employees feel like they're at one company. If you have to use method A to do something for one system, and method B to do something for another, then there will be a lot of time wasted bouncing back and forth.

There are a number of areas that I consider part of corporate IT that must be merged very early after acquisition:

- Email
- Corporate networks, including VPN
- Documentation systems
- Building access systems (i.e., not different keycards for different offices)
- Chat systems

IP address spaces collide, email groups must be re-formed, reporting structures change. The sooner these can be resolved, the better for everyone involved. With one acquisition, I saw a team proactively sit down with the network engineering team of the acquiring company and ask what IP ranges they could be allocated to make sure that there would not be address conflicts, which would make internal routing of traffic more difficult.

I witnessed one acquired company that was using some Internet-routable IP space they'd chosen arbitrarily for their internal networks! Soon after acquisition is a great time to prioritize that kind of clean up.

## Testing — Quality Engineering

Often during M&A, the testing infrastructure lapses compared to that of production. While there is a lot of attention paid to the product, revenue, and market share during the due diligence process, there is less attention paid to the systems that are used to develop that software. Some companies may even skimp on their testing infrastructure as a way of saving money to make the numbers look better.

Often a company may be in a state of transition between two different software delivery life cycles (SDLCs) at the time of acquisition. I worked with one portco that was using containers in its development and testing environments and rolling out all production deployments with old-fashioned tarballs. That violates a rule I had repeated to my SREs for years:

*Stage is like prod. Stage is like prod. Stage is like prod.*

Staging environments must be treated like production environments in most ways. This means that if we're deploying on Oracle databases in production, we can't substitute Postgres in staging because it's "cheaper." I talked to one quality manager who said her team spent 60 percent of its time debugging environmental issues and 40 percent on software. Staging (and all pre-production environments) should be *representative test environments*. The more closely they are representative, the more confidence we can have in the code that was tested. Skimping on testing will not be cheaper in the long run. It will result in lower quality and more frequent, longer outages.

If we're using infrastructure as code, this is much easier. Use the same code to build the staging environments that's used to build production. Like in all good code, we keep our configuration separate from our business logic, which makes sizing the environments appropriately a much easier task as well. If we have code that builds a Redis cache in production, we need to also build a Redis cache in staging, preferably with the same exact code, with perhaps smaller cache sizes in the staging environment. If we deploy our production software packaged up into tarballs, then the staging environment cannot be deployed using containers.

If one company has a Quality Engineering department while the other has a Quality Assurance one, it could be a big win for both companies. As we've discussed, quality engineering is about enabling developers to test quickly and easily. It's a huge improvement in velocity over throwing something over the wall to manual testers. The manual testers who are retained can be trained into quality engineers, and this department can be leveraged to improve outcomes for everyone.

## Monitoring

One place where pre-production environments often suffer, which results in complexity through sprawl in M&A, is monitoring. It can be *expensive* to monitor everything in all our environments. Many companies will attempt to skimp on the pre-production environments and save their monitoring budget for production. This is rarely a winning strategy. As we learned when discussing security, it's faster, cheaper, and easier to fix problems earlier in the process. If we skimp on the monitoring systems in pre-production environments, we are *increasing* our chances of deploying catchable problems out to our customers (where they are expensive to fix).

In fact, it's not as important that we monitor everything in our environments as it is to be able to use monitoring to give us answers. In the case of testing environments, this can mean things like performance. If we're testing a new release of the software and we discover that the CPU utilization increases by 90 percent, that's important to catch before we bring down the production environment.

> *The secret to monitoring is to be able to easily confirm or refute hypotheses.*

It's important to have base, high-level monitoring of our environments and applications. However, it's impossible (and prohibitively expensive) to monitor each and every possible scenario that may arise. When we're trying to explain the cause of some outcome, we're often in the realm of unknown-unknowns. We could wrap every method invocation with a counter,[43] but what would be the utility? Are the number of invocations of a method likely to be the cause of our next outage?

Instead, we need to make it easy for developers to instrument sections of the code so that they can answer hypotheses about how the code is behaving, because these developers have the most complete mental models of the code. This means that Operations needs to put the systems in place that make it possible for developers to consume the instrumentation, and it means that developers need to create internal libraries that make it easy to quickly obtain proof or refutation of hypotheses.

Often when multiple companies are combined, multiple half-implementations of monitoring systems are combined. This makes it incredibly difficult to share useful knowledge and experience of these monitoring systems within the organization. Some systems may be

---

43  https://web.archive.org/web/20141201114456/http://blog.librato.com:80/posts/2014/7/16/metrics-driven-development

designed as pull-based systems for operation in the relatively slow pace of a data center. Others may be push-based systems capable of emitting valuable monitoring information in the hundreds-of-milliseconds lifespan of a container running within a cloud provider's function-as-a-service platform.

Regardless of the style of the system, a well understood and supported monitoring system can be the difference between an extended outage, resulting in loss of customer trust and revenue, and a system designed to respond well in advance to leading indicators, providing the ability to expand dynamic capacity to meet imminent needs.

## Deployed Architectures

One area of wild variation between (and even within!) combined companies is the style of software deployment. If the company is large enough, there will be various styles, from on-premises (which we will address in the next chapter) to monoliths to modern containerized and function-as-a-service deployments. As we've discussed, the ability to deploy with quality and speed is a differentiator for the business, and quality and speed are much easier to do with modern deployment techniques. The modern tooling is opinionated in this direction. When looking at younger add-ons during due diligence I often see these modern techniques being used extensively.

Many private equity firms acquire established software companies with monolithic software architectures that have been deploying at customer sites or in their own data centers. It's much harder to deploy with speed and quality in these situations. There's typically an effort to break up the monolith into smaller services or even microservices.

With a monolith, there are myriad dependencies between the various software components. As opposed to a service-oriented architecture (loosely), these dependencies are typically tightly coupled. This means that for every release of the software of a certain size, there needs to be extensive (often manual) testing that must be completed to ensure that any single change doesn't have a negative effect on other parts of the software. This slows things down.

Because changes take a long time to develop and test, more changes are batched into these deployments. Instead of this extra testing making things safer, the releases of software become even more risky. Instead of deploying a few small or even single changes, like in the modern deployment methods, dozens, hundreds, or thousands of changes are released at once in a big batch deployment. Every additional change introduces more complexity, and it becomes more difficult to reason about the interactions between the numerous software components released at once. When there's a problem, it's equally difficult to choose which of the thousands of new interactions was the one responsible for the failure.

This was such a problem at Salesforce that they had an initiative with some of the most senior software architects in the company to break things apart into more manageable pieces. With its recent ventures into public cloud deployments, it would appear that this was a smashing success.

Many companies will set down the path of breaking off pieces of a monolith into services. However, they will often do so without the loose coupling required in a service-oriented architecture. When this happens, instead of being able to release each service individually, the services are batched up into a *suite* and deployed simultaneously. While there can certainly be an increase in development and testing speed in this scenario, deployment frequency remains the same. Each release is still a nail-biting affair of many changes released at once.

# When All You Have Is a Monolith ...

When I was at Salesforce a decade ago, they had a project called BUTC (pronounced butt-see, not kidding) which stood for Break Up the Core. Salesforce at the time was a big monolith and they knew they needed to break it apart.

A few years ago I was talking to a managing director at a big five accounting firm who told me that all their junior associates had somehow picked up the mantra: Microservices good, monolith bad.

So, what's the problem with monoliths? From the DevOps movement, we know that small batch sizes for releases create better outcomes than large. Monoliths create large batch sizes, implicit dependencies, tightly coupled components, etc.

## Wither Services

So, what's the alternative? Services. As microservices expert Chris Richardson explains,[44] a small monolith is fine, but as the size of the monolith increases, the ability to rapidly test, iterate, deploy, etc. decreases. I had a startup client that developed on a Node.js monolith. I recommended they package it up in a container, which would help encourage them to split out services, but I didn't recommend they destroy it.

---

44  https://chrisrichardson.net/post/microservices/devops/2020/11/24/microservices-enables-devops.html

I also had a client that had what *Team Topologies* calls "a distributed monolith," where the services were all tightly coupled and had to be released at once. Generally this happens because of a shared data store. However, a distributed monolith has many advantages over a regular monolith as long as we start versioning our APIs and aren't modifying the data layer. It can be released independently, tested, etc.

## Simplifying

Services allow us to embrace all that DevOps goodness and mature our (often very expensive) storage architectures at the same time. I've seen this happen in at least two different ways:

- *Intermediate layer for scalability*: I've seen places where the volume of traffic was so high that they had a Java layer with connection pools between their front-end code and the database that handled all the connection requests. Otherwise the database would be overwhelmed with constantly building up and tearing down connections.

- *Intermediate layer for flexibility*: The other patterns I've seen are those of a proxy layer, which passes requests directly to the database, or a layer that abstracts away the database entirely so the upstream services have no idea what's on the back end.

This second pattern is incredibly valuable for portfolio companies with a lot of acquisitions. By using this second pattern, I helped one company achieve a 70-percent storage cost reduction by migrating from one database platform to another. The entire time, the application itself was none the wiser.

## Good or Bad?

Nothing is good or bad. Each company is at the stage they're at. When a monolith is large enough, it can drag the engineering team down with it. However, there are many established and well-known ways to get portcos out of that debt and into the DevOps future.

## That Will Never Work Here

Often portfolio companies will explain that one of the reasons they batch up their changes into infrequent releases is that their customers won't tolerate more frequent deployments. Certainly for an on-premises product this is likely.

However, these companies are thinking about releasing *features*. At Salesforce they would make a big deal about how they would only have three software releases a year. In reality they would push bug fixes to production multiple times a day!

Going back to the *Project to Product* book that we discussed previously, Dr. Kersten describes four "flow items" that represent the work flowing through the value stream:

- *Features*: things that add functionality for customers. Typically the business logic, customer interfaces, etc.

- *Defects*: bugs. Errors in the software that yield incorrect results for the customer or the teams operating the software systems.

- *Risks*: security and compliance. Things that literally put the company at increased risk of data breaches, ransomware, etc., or failing a required audit.

- *Debt*: technical debt. The problems that accumulate over time that impact our quality and velocity to deliver products to our customers.

There are many important components that can be released frequently to improve the product, reduce costs, or increase performance that are not features, which can still have a major impact on improving the business. Customers have a high tolerance for these to be released frequently as long as they don't cause adverse situations. They rarely even know. I often ask: "If there is a major security hole that exposes customer data, do you think your customers would prefer you wait to fix it?"

## Datastores

Alas, our poor data management team that opened this chapter has a different problem — proliferation of datastores. Why? Why are there so many database technologies after a number of acquisitions?

*The First Rule of SaaS: Don't lose customer data.*

Customer data is stored in the datastores, and one of the most proven ways to not lose customer data is to touch customer data as little as possible. Most CTOs came up as software developers. They have a lot of opinions on programming languages, web frameworks, etc. There are certainly some who will have opinions on graph databases, NoSQL, etc. Most know to do everything in their power to prevent or minimize the loss of customer data, so making major changes to the acquired datastores is often avoided in favor of other areas.

The datastore teams are often left attempting to support multiple systems of record. There's a reason *Team Topologies* calls these teams "complicated subsystem teams" — it's complicated! Modern databases are complex beasts requiring in-depth knowledge about indices, B-trees, caches, I/O performance, etc. Knowing that many database technologies at a deep level is next to impossible. We would like to achieve maximum performance without relying on vertical scaling until that's no longer an option.

What often happens is that a single expert is brought in (or retained) for each database technology, which reduces collaboration, increases burnout, etc., as we've discussed. If we're able to reduce the number of database technologies, even if we don't reduce the number of datastores, the teams can gain efficiencies with tooling, tuning, and the like. The teams can learn from and support each other and better support their (internal) customers. Complexity isn't free! When they are able to work more closely with their customers, we get the reinforcing part of the loop.

**DevOps Patterns for Private Equity**

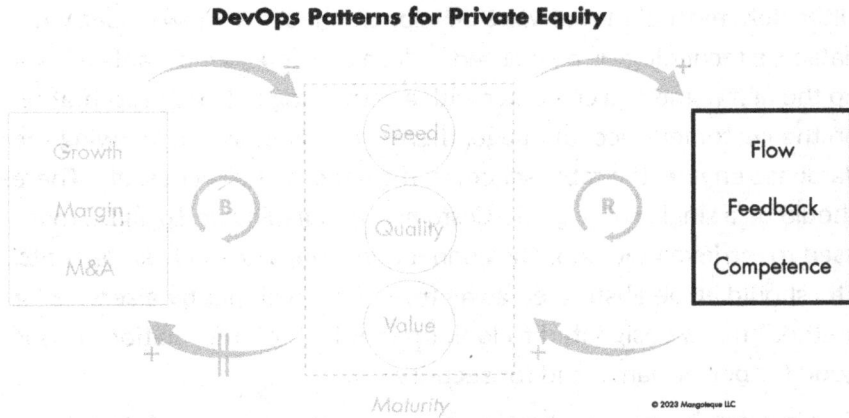

Database teams can have toil too. A database team lead explained to me that her team was spending 15 percent of its time running procedures that could be handled in the application code. She was working with the development teams to get that time back so her team could work on things that would bring more value to the business. The development teams needed the ability to ship with speed and quality to be able to take on that work alongside their regular product commitments. The reinforcing loop.

Consolidating around a few well-supported database technologies helps teams really understand the things they're responsible for and reduces complexity.

## Reducing Complexity

With many problems in DevOps, the solution is to reduce complexity. Not simplifying the datastores themselves, but simplifying the variety of datastores. We like to say in distributed systems that you don't need to make it complicated, you get the complexity for free! More expertise in fewer database technologies is better than less expertise in more. For some databases, migrating to a similar technology is rather straightforward. For others it can be more of a challenge.

Ultimately, most of the software developers shouldn't know or care what database technology is being used. If I'm a developer who wants to look up the first name of a customer with a certain login ID for presentation on the customer's account page, there's no advantage to knowing the database engine, the table schemas, the indices, the caches, etc. There should be a single call (e.g., *GetCustomerAttribute(Fname, LoginID)*) that's used to retrieve that data. The connection strings, the SQL statements, etc. should all be abstracted away from the developer by a service (or method) that's designated to look up that kind of information. This is good for performance and for security.

This pattern is good for helping to reduce the number of datastores necessary for the database management team to support and understand at a deep level. It's possible to migrate to a completely different database engine without the front-end engineers even knowing! To them, they're making the exact same calls in their code. The complexity is handled by the development team, which works with database management to return the appropriate data regardless of where or how it happens to be stored in the back end.

Is this trivial? Of course not. However, having a small internal product team that handles the complexity required to interact with customer data *is* trivial in comparison to teaching hundreds or thousands of engineers the intricacies of, and granting the permission for, interacting directly with customer data and risk violating the First Rule of SaaS.

A portfolio company had a robust and highly scalable performant datastore. They were doing high-volume work and needed to be able to store large amounts of customer data arriving continuously. Because it was performant, the datastore was able to keep up with the sustained load, but because it had to store large amounts of customer data, it was expensive.

The team had already abstracted the datastore from the engineers who consumed it. Despite being a large, complex data storage solution, the consuming engineers only needed to make API calls (through well-established methods) to retrieve and store the data necessary. Having all the engineers become proficient in the intricacies of the database technology was of no advantage to the business.

The team did extensive research on ways to make database storage more efficient while maintaining the necessary performance and durability guarantees. Over the next few months they introduced a caching layer that could reduce the load on the database.

Over the following few months, they introduced compression into the data storage so that each byte written would be more efficient and take up less space. Over the next few months, they introduced clustering technology and permanent storage at the caching layer. Over the following few months they then tested and measured. The more capable the new storage tier became, the less money was necessary to operate the original technology.

At each stage of the process new capabilities were released behind feature flags and dark launched. The engineers would release code in the morning and then activate the feature flags. During the course of the day, the output of this dark-launched code would be checked against the existing system to see if the results matched exactly. If not, there was more work to be done. If so, it was tested for performance with p99 measurements, etc. At the end of each day, before going home, the engineers would turn the flags off, ready for another round of testing the next day.

Eventually, after approximately 18 months, the team had completely replaced the original storage technology with a new one. Storing customer data had been this portfolio company's greatest expense, and they conservatively reduced their storage costs by a whopping 70 percent!

This entire data storage tier was replaced without sacrificing durability, operability, or performance and required *no code changes by the consuming engineers.* Those engineers continued to make calls to the service that handled retrieving the data, and those working on the back-end storage layer handled all the complexity of where to retrieve what data, when.

## Cultural Obstacles

*"Culture eats strategy for breakfast."*
– Peter Drucker

As most experienced engineering leaders know, the technical problems can be the easiest to solve. If we instruct a computer what to do, it will generally do the same thing each time. We've discussed that humans have different life experiences, and different communication styles, have fights with their partners, and forget to have coffee. With the same human, we can get different results in similar situations. Biology is messy.

When we take the culture of multiple companies and attempt to combine them into one, problems can become multiplicative. Instead of three datastores we have six. Instead of Site Reliability Engineering (SRE) being advisors, they're on-call. Instead of quality engineers, we have manual testers. Which path do we choose? Who has the right culture?

Of course, no one has the "right" culture, and often the solution is to take the best parts from each. Activities like the architecture meeting (from the SRE meeting plan) can help. Maybe these conflicting choices are like a form of technical debt, but like technical debt they're not engineering problems, they're *engineering leadership (business)* problems. They can only be fixed by leadership creating changes in the system.

*"I should estimate that in my experience most troubles and most possibilities for improvement add up to the proportions something like this: 94% belongs to the system (responsibility of management), 6% special."*
– W. Edwards Deming

### Vestigial Problems Are Amplified

One concern is that, while these problems may have existed in the original companies, when the companies are combined, the effects are amplified. One portfolio company decided in 2018 to abandon having a formal quality engineering practice. Because it had been the practice of the company for many years to have an emphasis on QE, it was part of the culture. Performing these practices was just a part of the way they worked and things seemed fine.

After the company was combined with a few other companies that didn't have this practice, the results were a disjointed mess. Some teams had quality practices, others did not. Sometimes engineers got together to talk about how to improve quality, and other times they were simply too busy. Quality practices at the larger company became a best effort endeavor. If it was important to your team, then it probably happened.

Other teams just carried on not having to worry about testing, or quality. An essential part of delivering software with speed and quality had become simply an afterthought.

Other functional areas like performance testing, security, and Agile practices are especially susceptible to this amplification during buy-and-build.

## When Work Isn't Visible

As we discussed in the chapter about managing work, making work visible is essential for a high-performing organization. When multiple organizations are combined, this can be a challenge.

Each company is likely to bring its own ticketing system. When they're combined, if they're lucky they're using the same ticketing software. Otherwise, the work of one part of the organization may be hidden from another. This may be less of a problem for individual teams, but for leadership who are trying to understand the flow of work through the system and what milestones are being achieved on different initiatives, this can be a major impediment to managing the organization.

Chat can have similar problems. Even when they're combined into the same chat system, if there are silos, people are inclined to have private chats. They want to tightly control their interactions with other areas.

If discussions are held in private instead of making work visible, it's difficult:

- to bring the collective knowledge of the organization to bear on problems.

- for members of other teams to drop into a chat room and ask a clarifying question or inform a team about an upcoming change.

- to see what changes have been made or which components are failing.

- for teams to make decisions that will allow them to meet other teams seamlessly at a future state.

This privacy pattern creates silos of information, as opposed to Westrum generative culture, in which high cooperation and messaging are encouraged — impossible to achieve when the primary methods of open communication are effectively locked down.

## Data Centers

One might suggest that a discussion of data centers belongs in the technical, not cultural, part of this chapter. But choosing to run in a data center vs. cloud is a cultural decision about how the company will deliver.

It *is* possible to deliver with speed and quality in a data center environment. Unfortunately, most companies aren't very good at it because it's a hard thing to do well. Most data center build-outs are fragmented affairs, where each group (network, systems, and application engineering) has a part to do, but there's never a cohesive approach to installing, allocating, and retiring equipment, configurations, etc. as a holistic life cycle. Nonetheless, people often insist that a physical data center is necessary. Generally the reasons come down to one or two main arguments:

- A data center is more secure than the cloud.
- A data center is cheaper than the cloud.

Upon closer examination, it becomes evident that these are simply matters of implementation rather than hard and fast rules. I've seen countless data center environments with the M&M security model (hard candy shell, soft in the center). The network administrators put up firewalls and intrusion detection and bray about how secure everything is because they passed a penetration test. Meanwhile, an attacker exploits a vulnerability in the code and is soon able to move easily around within the network. Security is neither easier nor harder in the cloud compared to data centers. It comes down to implementation.

What the cloud *can* do is expose things as an interface (API) so that security checks, audits, etc. can be more easily performed. Want to know the permissions on an object store bucket in the cloud? Pretty easy. Want to know which machines are mounting an NFS store inside a data center and compare that with what should be mounting that share? That's quite a bit more work.

Arguments about costs are a distraction to the discussion of data center vs. cloud. It's not up to a cloud provider to provide users with cheaper hardware, and most cost calculations are massively incomplete. As we'll see in the next chapter, most data center implementations use a single data center, sometimes two (for redundancy). But it's actually more redundant and cheaper to use three. For those running their own data centers, it can be hard to do and expensive. If the company has the staff to maintain all that's needed in the data center properly, then it's probably no longer saving money compared to using the cloud. (Amazon is better at running data centers.)

"The cloud" is something that's talked about often. It's something that can increase the multiples for a portfolio company during their exit. It's something on-prem software companies move to. It's something that allows the most innovative companies in the world to be innovative. It's also something many people don't understand in the right way — why it's important and what makes it so different.

Cloud is commodity-based computing. But that doesn't make it cheap. Cloud is manageable with API calls and code. But that doesn't make it easy.

> *The advantage of cloud is flexibility, not cost.*
> *Massive cost savings can flow from flexibility*
> *but that comes from engineering, not the cloud vendor.*

Whether you run your software in a data center or run it in the cloud, there's one main secret to running it well: In either environment, you should be able to take any server and destroy it, and if you can get an exact replica of that server in place in a few minutes, then you're doing it correctly. The industry has been moving toward ephemeral infrastructure like containers and function-as-a-service or "cloud native." Coincidence? Not at all. Those archetypes allow you to deploy in the same paradigm as destroying a server.

Everything flows from that ability: where data is stored; how the networking, containers, security, permissions, access control, resilience, etc. are all set up. If customers don't notice and it's trivial (minutes), then that also means that *new* storage paradigms (for cost savings), *new* services (for product innovation), and resilient infrastructure (for customer retention) are all easy for that company.

### DevOps Patterns for Private Equity

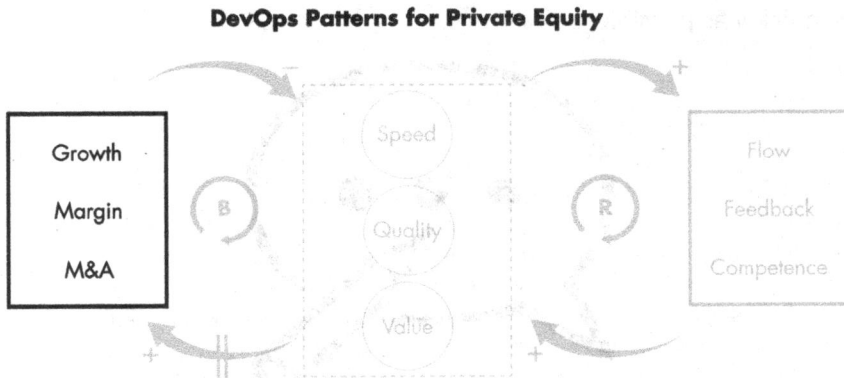

© 2023 Mangalaque LLC

When cost savings, innovation, and delighting customers are easier to do, then below-the-line costs go down, and above-the-line numbers go up. The cloud vendor can't make your engineering organization high performing, it can't reduce your costs because it's providing you a ton of value, but when companies can leverage the secret to cloud, they are only constrained by their innovation and imagination, not by their execution.

One of the biggest differences in culture between a data center and the cloud is that data center installations are long-lived. Companies test different hardware from vendors. They make large capital expenditures. They design something that will last for years. In the cloud, if you want to reconfigure the network, or destroy and build a new one, you can do that multiple times in a day.

As we learned at the beginning of our discussion, in today's business environment, it's the fast beating the slow. If you build out your data center environment and have the ability to dynamically allocate hardware, then you can move quickly. When you can provision that infrastructure using infrastructure as code, then you can move very quickly. Even the market-leading hardware virtualization software companies have extremely awkward interfaces for managing the infrastructure programmatically. They invest most of their resources in creating fancy and powerful GUIs. But companies that want to move fast don't want fancy GUIs or consoles, they want to be able to write code that creates self-service interfaces that allow teams to innovate as quickly as possible.

"Dave, I was talking to one of our portfolio CEOs, and she has major concerns about her cloud engineering team."

I looked up the company and looked at the skill set of its cloud/SRE team.

"Yeah, I would have major concerns as well. Hardly any of them can code."

Today, a cloud operations team that's not made up of coders is going to be a low-performing team. They need to understand infrastructure and know how to code. Developers write code for business logic. Operations folks write code for infrastructure.

*"SRE is what you get when you treat operations as if it's a software problem."[45]*
— Google SRE

Private equity firms work with companies to take them to the next level. Cloud (or advance data center implementations) require code to deliver repeatable results at speed.

Terraform is great, but it's not a programming language, yet. Ansible is interesting, but when you want to do anything advanced, you need Python. If I ask the Amazon Web Services API to return all the hosts in a specific region, located in a specific availability zone, with a specific tag, I'm going to get a multilevel data structure. To parse it, I'm going to need a programming language. Command line hacking is not enough.

If you want to know why a particular company has so much trouble scaling, look in cloud engineering for programming languages. That's where you'll find the true high performers.

The vast majority of companies that choose to deploy their hardware in a data center instead of the cloud are making a cultural decision to go slow. They're making a choice to do a large amount of undifferentiated heavy lifting. They're making a choice to rely on vendor magic to protect them from failure or help them to recover if they do fail. If a product needs a very large amount of storage, this can be a fair business trade-off. If a company is doing it because it's more secure or cheaper, the competition in the cloud will most likely be just fine.

---

45  https://sre.google/

# That's MTBF Thinking

*"This would be so much easier if we could just use VMWare Motion."*

The words hung in the air after I heard them. I was on a call with a number of engineering leaders at a portfolio company that was having stability issues in its production environment. The company, and its customers, were frustrated over the number of incidents they'd been experiencing recently, and we hadn't even gotten into the issue of how to scale the platform by 10x. Without changing the way they approached the problem, they would never really solve the stability or scalability challenges. I recognized it immediately: it was MTBF thinking.

## MTBF vs. TTR

I began my career in an industry that was focused on MTBF (Mean Time Between Failure) thinking. I got pretty good at Veritas Cluster Server. I built cold standbys and tested their failover. I worked for companies with mature disaster recovery and business continuity plans. I also got pretty good at VMWare, even coding against its Perl SDK. By that time, however, I'd embraced TTR (*minimizing* Time to Recover) thinking.

MTBF is great for disk drives. Not for production SaaS services. In service delivery, especially with complex distributed systems, we need to be comfortable with the fact that failures happen. Once we're comfortable with that fact, we can begin to operate as if those failures are expected, because they are. Artificially trying to suppress entropy is folly. Jez Humble explained it well back in 2013 when quoting Nasim Taleb's *Antifragile* book:[46]

> *The problem with artificially suppressed volatility is not just that the system tends to become extremely fragile; it is that, at the same time, it exhibits no visible risks.... These artificially constrained systems become prone to Black Swans. Such environments eventually experience massive blowups ... catching everyone off guard and undoing years of stability or, in almost all cases, ending up far worse than they were in their initial volatile state.*

Without dwelling on whether antifragility is appropriate for distributed systems, the Black Swan events are real. Teams that try to artificially suppress entropy will be unprepared to deal with events as they happen. That's why Netflix invented the Chaos Monkey. That's why I teach portfolio company teams to practice Gameday and GoLive (Production Readiness Review) exercises. We need to prepare *for* failures, they *will* happen.

VMWare Motion is MTBF thinking. It promotes the idea that we should strive to keep systems running as long as possible, to try to suppress volatility and create the illusion of stability. This is bad for security, stability, and scalability. The secret to cloud is to be able to take any server and destroy it, and get an *exact* replica of that server in place in a few minutes.

## The DIE Model

How should we operate our systems instead? Sounil Yu of JupiterOne popularized the idea of the DIE model when talking about cybersecurity,[47] but it's also a great way to think about modern architectures beyond that specific realm.

DIE stands for Distributed, Immutable, Ephemeral. This is the opposite of MTBF thinking. We are not trying to keep everything running all the

---

46  https://continuousdelivery.com/2013/01/on-antifragility-in-systems-and-organizational-architecture/
47  https://www.cisa.gov/sites/default/files/publications/Daily_Keynote_with_Sounil_Yu_508pobs.pdf

time. We are not trying to use VMWare Motion to keep the machine image "alive" even as the hardware underneath changes. Instead, we build systems to be distributed to minimize the blast radius of any one individual failure. Our systems are immutable, so that instead of modifying a system over and over and trying to perpetuate its existence, when we need a new one, we build a new one, and that process is easy. Finally, considering our systems to be ephemeral emphasizes quick replacement of lost resources, understanding that entropy is real, and that the best way to ensure availability is to recover quickly, rather than trying to futilely prevent failure. It's Cattle vs. Pets.

So how does this address the main problems of our portfolio company above?

- *Stability*: No longer are we worried about failure or preventing failure. Failure will happen. But instead of a failure being an all-hands-on-deck firefight that celebrates heroic efforts, failure is a run-of-the-mill activity that's either remediated by automated systems or, if the company is not yet at that maturity level, results in a nonchalant, simple replacement of the failed resource by a human operator.

- *Scalability*: Along with techniques like feature flagging and dark launching, the ability to scale a platform is done through a series of experiments to understand how a system scales, where its bottlenecks are, etc. By following the DIE model, we naturally create a collection of smaller horizontally scalable services, while also giving ourselves the ability to quickly and safely iterate on hypotheses and determine how to move bottlenecks to more scalable parts of the system.

# Go DIE

Failure is a learning experience in life, and it's one of the ways that we come to form more complete mental models of our systems. Once you come to recognize the artificial suppression of failure (MTBF thinking), it jumps out at you like a big red flag. Moving away from MTBF thinking has given us robust distributed file systems, Kubernetes, S3, and a host of other things that power the modern world.

Learn to embrace TTR thinking. Design your systems for DIE. Give your teams the ability to deliver with security, stability, and scalability. Recognize MTBF thinking, and help it die.

## Keep the Developers Moving as Fast as Possible

I used to tell my SREs they had two primary responsibilities:

1. Keep the site up.
2. Keep the developers moving as fast as possible.

From technical debt to monoliths, from change advisory boards to data centers, often what we find when merging technical organizations is drag. Drag on the ability to deliver. Drag on the ability to meet customer expectations. Drag on the ability to meet our deadlines and deliver as promised and on time.

We can combat a lot of these missed deliverables by using Agile techniques combined with making work visible as we've discussed. But sometimes the outcome of making work visible (like when examining value streams) is the ability to see how much drag is affecting the system, which can be in the form of various kinds of technical debt, or it can be in cultural debt when the teams are still exercising old ways of working.

Regardless of how we reorganize the teams or make work visible, one of the most effective ways to keep teams moving quickly is to make operations work self-service as much as possible. We saw this when talking about platform teams. We saw this when talking about how operations provides a platform, and that the best service is no service. We saw this in our discussions of shifting left, about making it easy and safe for teams to perform work earlier in the process.

I've seen some incredible engineering work made possible because functional teams allowed developers to deliver quickly. We've done Service Delivery Assessments of some companies that would allow developers to define what infrastructure was required through simple YAML checked into their source code repositories. However, there was one team that stood out because they took it a step further.

This team allowed Terraform to be embedded directly in the YAML file. If there was a brand new cloud service that was supported in Terraform, but not yet supported by the SRE team's custom infrastructure definition tools, the developers could still write raw Terraform to be able to experiment with the new cloud service in the interim. We were very impressed.

It's this commitment to ensuring that there is little wait (waste in Lean) in the system, whether through infrastructure definitions, CI/CD, or in shifting left, that will help the team to deliver quickly and as promised, because they're rarely blocked by waiting. Even without embedded systems engineers, the teams become effectively cross-functional teams because they have everything they need to be able to deliver on their own.

This ability to control the development, deployment, and possible remediation efforts allows teams to keep their deliverable promises.

When teams aren't able to deliver their own software, when they're not in control of their delivery cadence, delivery becomes harder. There are many portfolio companies where the work is held up in hand offs between silos. When companies have been formed by buy-and-build, there are multiple areas of the company where the siloed delivery mechanism can be present.

These situations cause the company to miss deliverables. They cause companies to have extended and frequent outages because the people who wrote the software are no longer responsible for it. Those charged with operating the software scarcely know what changes have been recently made and are generally only able to recover from infrastructure outages if they've seen that exact problem before. A problem in the software itself either needs to be rolled back if possible, or needs a knowledgeable developer to be found.

This leads to irate customers who, if they're of sufficient size, will receive repeated apologies from the CEO. One portfolio company talked of having burned all their goodwill with their customers by having repeated and extended outages. Of course, when there's no more goodwill to burn, there are either severe SLA penalties that are ruinous to top-line portfolio company numbers or the customers churn out, which is equally damaging.

This can be one of the reasons that companies choose to move their installations out of the data center and into the cloud. The hope is that by moving to an environment built to be managed strictly through interfaces, they will be more able to empower teams to rapidly deliver their own software, whether by leveraging managed cloud services, writing their own interfaces, or some combination of the two.

# From On-Prem to SaaS

*"Uncontrolled variation is the enemy of quality."*
– W. Edwards Deming

"We'd like you to help us migrate five products to the cloud."

"OK," I said. "What's your timeline?"

"Six months," came the reply.

Six months to migrate five products to the cloud for a company that had no experience with cloud architecture or operation. In the ensuing emails I used words like "bold" and "ambitious" but, at the same time, I knew it was completely unrealistic.

Sure, it was possible to just stand up infrastructure that looked similar to what was in the data center, but the cloud is a different paradigm. Done properly, it can be a strategic differentiator for the business, enabling it to do things never before possible. In the end, that was exactly the case for these folks.

"Well, I doubt if you'll need me for all five products. Maybe for the first three, but after that, your team will get the hang of what's involved."

"Oh, we think we'll need you for all five," came the reply.

Over the next few months, we reorganized the team into functional areas. We tried multiple infrastructure-as-code tools until we found the right one. We set up monitoring. We tested failure scenarios and improved our ability to detect and respond. The team learned what it was like to truly operate environments on a cloud platform. We moved over one product.

We began work on another. It took almost three years to migrate all five products. I helped the team for one and half products before they were off and running by themselves.

# SaaS as a Business

A former boss explained to me that one of the best business models is what he calls a Spreadsheet in the Cloud. Find a problem that people have solved with some complicated formula in a spreadsheet, re-implement the formula as code on a website, and store the data in a database. Soon you're solving a real customer problem in a way that scales relatively easily.

It's a great business because, once it's working, the marginal cost to add additional customers is negligible. Databases can handle millions of rows before they need to be scaled up, and much of the hard work is handled by a cloud provider. The code and infrastructure are highly leveraged, and as long as the lifetime value of the customer is more than the cost to acquire a customer, it's a money-printing machine.

When private equity firms choose to move an on-premises (on-prem) software company into a SaaS company, the process isn't so simple. But the business advantages of running software that's already solving real customer problems in a way that scales much more easily, and with much better margins, than on-prem are compelling.

A lot of the patterns we've discussed can be brought to bear during this transition. When helping companies on this journey, like the one above, I think about the process as categorized by three major phases: on-prem, hosted, and SaaS.

# On-Prem

The software has been performing well, customers love it, and they've been using it for years. It's a very desirable situation. There's one thing this installation type cannot escape: complexity. When software is deployed on-prem, the company isn't managing a software installation. They're managing as many installations as they have customers: 1,000 customers? 1,000 unique installations to manage. This is an expensive way to run software. If the software is deployed on multiple machines (virtual or physical), then it's responsible for 1,000 times that many machines.

Any upgrades or maintenance must be coordinated with each customer, even if it's that extremely rare on-prem piece of software that can be upgraded in place with no downtime.

- When staff perform an upgrade or maintenance, they must determine the current status of the environment. For instance, when the customer is supposed to have two VMs for redundancy, they may have turned one off to use the resources for another machine. VM machine size changes, networking changes, and IP readdressing are all real possibilities. But the upgrade itself was only ever tested under ideal conditions, so danger awaits. I've worked with portcos that had engineers dedicated to doing nothing but manual customer upgrades, every day.

- These problems are exacerbated by time pressure when there's a problem at the customer site and company staff have to figure out what's broken. As Winston Churchill said: *"It is a riddle, wrapped in a mystery, inside an enigma."* The engineer knows they're walking into a trap. The problem could be with the on-premise software, it could be temporary, it could be a configuration change. Often on-prem customers are reluctant to accept upgrades, so it could be a very old version of the software that behaves much differently from the software on newer installations. It's extremely manual, tedious work.

- In the best-case scenario, failover is multiple machines in a cluster. Failing over to another data center is rare enough to be almost non-existent. Almost every outage is a customer-affecting outage, putting more pressure on the assigned engineer to diagnose and fix the problem. As anyone who's responded to an outage knows, additional pressure never helps to get the situation resolved faster.

- Because this is a local customer deployment, there are often custom solutions (both hardware and software) implemented. None of these customizations will affect other customers, so what could be the harm, right? (We'll find out later.) Often these are additional columns added to the database to facilitate custom reporting. I worked with a portco that described its account managers being more like customization consultants than customer relationship managers. It was a series of snowflakes.

# Hosted

Quality Internet connectivity has become ubiquitous for businesses, and cloud security concerns have diminished, so the need to run on-prem has greatly diminished over time. As we've seen, those installations tend to be extremely manual and resistant to automation because of customization, and thus they're expensive to maintain. After acquisition by private equity, the board decides to move these installations from the customer site (on-prem) to the cloud. I call this stage of leveling up "hosted."

The company is still managing individual installations for each customer, but mercifully (for its staff), customers have much less ability to modify their installation at this stage. The engineers build and operate the software and the customer is simply presented with a login screen. Often this address is customized for the specific customer because it refers to a specific installation. The engineers also have much more control over when upgrades happen because the environments are more under their control.

Deming, the father of quality, used to talk about the difference between what he called common cause and special cause variation. In the context of the cloud, common cause variations are due to the nature of being on a cloud provider. Machines will disappear, networks will partition and come back. These are the expected variations that need to be handled as part of normal operation (TTR thinking). Special cause variation comes from unusual circumstances. Highly customized installations, database schemas for reporting, software configurations, etc. are unusual circumstances in a SaaS business and should not be handled but rather eliminated.

In this stage we will apply many of our DevOps patterns. In order to rapidly evolve the architecture, we will focus on flow. In order to apply what we learn after each experiment, we will shorten and amplify feedback loops. In order to have confidence in the operation of our new SaaS architecture, we will focus on competence.

**DevOps Patterns for Private Equity**

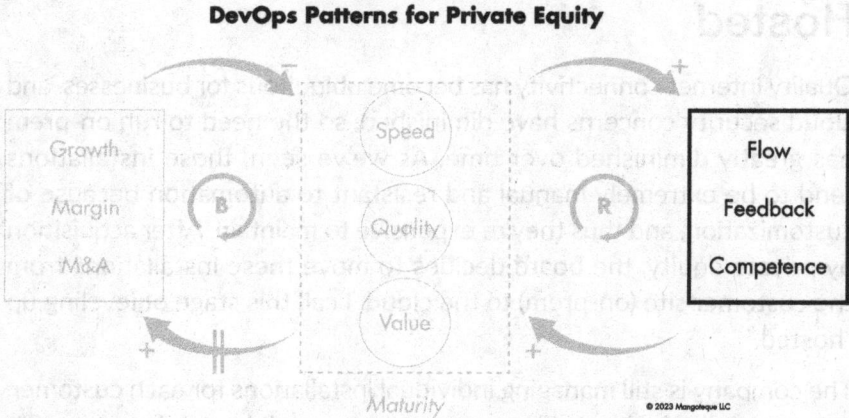

© 2023 Mangaleque LLC

# Feature Flagging

This stage of the journey will necessitate the ability to deploy software behind feature flags. That is, the ability to wrap code in such a way that this part of the code can be enabled or disabled via an external mechanism. We saw this when engineers replaced an underlying storage subsystem.

Robust feature flagging frameworks will enable us to expose software to individuals, customers, groups of individuals or customers, percentages of users, etc. Many software companies release *every* piece of software behind feature flags. This was common practice at Etsy, so that if there were operational problems in the middle of the night, the engineers had the ability to instantly turn features off.

Every high-performing software organization releases software behind feature flags to avoid the necessity of performing yet another release in the event of problems. Hygiene surrounding already-deployed flags cannot be overlooked.

# Dark Launching

Dark launching refers to the practice of releasing test software that has no effect on the customer. These can be things like algorithmic improvements for efficiency or cost savings, alternative storage systems (like we learned about in our 70-percent cost reduction scenario), or testing tax calculations with live data.

It's an extremely useful technique for any improvement to an existing system where we need to ensure that the customer's perceived behavior of both the old and new systems are identical or can meet expectations. One of the most famous stories in the industry was the release of Facebook Messenger. Each regular Facebook page carried additional code to test a messaging infrastructure. This ran for months of testing before any customer visible capabilities were displayed. By the time the software was formally released, the team had a high degree of confidence that the system had the capacity to perform as intended.

## Hosted Execution

The most common and biggest mistake I see companies make at the hosted stage is to just get running in the cloud with the idea that they'll worry about all the messy parts of operating in the cloud later. *The cloud is not just someone else's data center.* As we've repeated many times, the cloud is a different way of working entirely.

- Cloud deployments *must* be built using infrastructure as code (flow) and have telemetry built in (feedback). Thankfully most cloud providers have built-in telemetry. *This is non-negotiable.* Moving to the cloud is exciting, but the advantage of cloud is speed of iteration (not cost). If environments are created by clicking around in the cloud web console, the company will be both slow and in the dark. They have little ability to iterate quickly and are in an environment where hardware is ephemeral, with no way of knowing if something disappears (common cause variation). I was asked by a company to evaluate its architecture to solve some of these problems, and I had to tell them that there was no easy fix. You have to start over with a new system.

- At this stage, the main goal is to stop doing special-case things (special cause variation as discussed above). Differences between customer installations must be normalized or eliminated. In essence, if a company has 1,000 installations, they're trying to get to a point where they're operating 1,000 of the exact same thing, not 1,000 different things. Companies need to use the capabilities of the cloud to rapidly iterate on these changes and make the environment easier to operate and manage.

- This is the stage where a lot of hard work needs to go into how to deal with the fact that each customer is deployed to its own individual database. I had a client who was very proud of the fact that Microsoft Azure was running its 1,000 customer databases. I explained, "That's great, but you have 1,000 databases and you're not Facebook." Complexity is not free. If a company has added columns to a database schema to facilitate reporting and we run a database migration script as part of an upgrade, we may lose those columns, or the upgrade script may crash. We have a very real chance to violate the First Rule of SaaS. To achieve the best cost model with SaaS, the environment must be simplified. You don't have to try, you get the complexity for free.

- This is also the stage to start taking advantage of services that the cloud provider offers. This is a relatively easy way to bring more resilience into the architecture. Caches, databases (like above), queuing systems, etc. are all available on every major provider. I worked with a company that was still running a separate database for each customer inside a virtual machine while hosted in the cloud. They would have saved a ton of money, complexity, and aggravation by paying the cloud provider for that capability. This is also an excellent time to move any stateful data from the application servers into a persistent object store. This can be uploaded files, session data, etc. Since the servers are no longer holding customer data, they transition to cattle from pets. Instead of needing to be lovingly taken care of, it's the ability to get an exact replica of that server in a few minutes.

- Because the cloud offers almost everything as an interface (API), we can also start taking advantage of those interfaces. I worked with a company that was migrating from one cloud networking paradigm to another and wrote software that detected incorrect network configurations. This helped ensure, during the course of the migration, that there was a high degree of quality in each step of the process and that nothing was left in a state where it could cause a customer outage.

- Since everything is an interface, this is also a great time to start shifting left as discussed in the chapter on security. We can empower the engineering teams to a degree rarely possible when software was deployed on-prem. This will allow us to make changes more quickly and thus more quickly evolve the architecture (flow). The cloud provider console cannot give this, infrastructure as code is required.

- Because the cloud providers have monitoring built in, there is less guessing and speculation about the state of the infrastructure. This can not only help to ensure higher uptime in terms of the product itself but can also be a boon to capacity planning. Over time, the team should have a better understanding of what capacity is needed to operate the evolved infrastructure. The system should become easier to operate and more efficient, thus reducing costs.

- As different components of the software are deployed into the cloud, this is an excellent time to get in the practice of destructive testing (competence). I've worked with companies where no new services are allowed to be deployed without a joint Gameday exercise, where individual components are identified with their expected and observed failure scenarios, as we discussed in the security chapter (see Appendix A for a Gameday template). Those services are not allowed to be put into production until the two scenarios match. Because "repetition and practice lead to mastery," when we practice destructive testing, we can help to develop mastery in operating this software.

- When the company has successfully developed the ability to quickly and repeatedly deploy new customer installations, then they can start taking advantage of things like autoscaling for more resilience. As variation is reduced, it's easier to automate things because they're operating against consistent infrastructure. Instead of requiring a high amount of customization by dedicated account managers, the team can give its sales folks the ability to deploy customer installations after they've made a sale or allow customers to sign up themselves!

# The Hosting Site Rule of 3

Many people have heard of the SaaS Rule of 40.[48] But what about the Hosting Site Rule of 3? You probably haven't heard of it because I just made it up. I talk with a lot of enterprise executives about their companies, especially private equity portfolio companies. They're often working with software that has been successful for a long time, has achieved product market fit, and was almost certainly designed before the explosion of cloud. I often state, to shocked surprise, that of course:

*It's cheaper to run in three data centers than in two.*

That's impossible? Right? In what universe is three less than two?

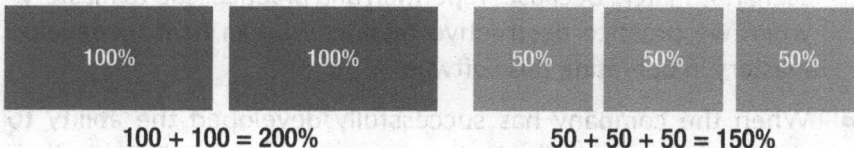

| 100% | 100% | | 50% | 50% | 50% |
|------|------|--|-----|-----|-----|
| 100 + 100 = 200% | | | 50 + 50 + 50 = 150% | | |

Most of those applications were developed at a time when the way to handle failures in an environment was to have a "disaster recovery" site.

---

48  https://www.scalevp.com/blog/a-quick-primer-on-the-rule-of-40

That meant that the primary site would handle all the traffic, and if there was ever an adverse event, the site would fail over to the secondary site. There were and still are a number of problems with this scenario:

1.  You'd need to purchase 200 percent of capacity, with the secondary site sitting idle, doing absolutely nothing. The secondary site needs to be fully built out to handle the full website traffic, but often just sits for months or years – *an incredible waste of capital*.

2.  Some secondary sites never bother to do the full build-out and spend all that capital (understandably). The company hopes they'll get their primary site back up before things get "too bad."

3.  Many secondary sites never get around to testing the failover or fail to simulate the correct conditions. When it comes time to use their secondary site, companies sometimes encounter a *strange loop* (the secondary site requires the primary site to be active to perform the failover because that had been the situation when they'd tested it!). This actually happened at Salesforce.

By contrast:

1.  If we run active-active-active in three locations, we only need to purchase 150 percent of the site capacity. All resources are always used. Keep this in mind the next time someone tells you how expensive the cloud is.

2.  We even have a measure of headroom to accommodate burst traffic if it arrives, and that can give us advance warning to deploy additional resources if necessary (or automatically).

3.  We can lose an entire data center and it's not a panic event. The site keeps working without strain and we can either spin up more resources in the existing sites, or work on bringing back the third site.

4.  The networking is vastly simplified. In AWS, if you deploy across three availability zones, the load balancers are deployed that way as well.[49]

Of course, it's possible to deploy the two sites as active-active, even if it's rarely done. But the costs are still 50 percent more. For architectures that

49  https://web.archive.org/web/20160315090624/http:/blog.librato.com/posts/metrics-everywhere

are making the transition, I generally work with the engineering teams on moving session data from the hosts to a cache and the local data files to an object store. This way, the cloud provider handles many of the HA features, leaving the application architecture concerns to the portfolio company engineering team.

I remember that when I learned this long ago, I actually felt foolish that I hadn't reasoned it out myself. How are your sites deployed?

# SaaS

The company has eliminated special cause variation between the different installations. It has taken advantage of cloud services. It may even have developed a database schema that will allow for multiple customers to be stored in the same database installation (multi-tenant). Its application has been augmented to support sessions as a way of distinguishing customer transactions instead of requiring a specific installation. This way, customers log in at *www.service.com* instead of *customer.service.com*.

- There's a marginal cost to adding customers. They can either sign up themselves (hopefully) or be quickly and easily on-boarded by a sales representative.

- Because we're not running separate software and infrastructure for each customer, both are now highly leveraged. A software deployment works for all customers, and there's a marginal cost to add infrastructure as the business scales.

- The company has a mature deployment and operating model for the software installation. It can release changes behind feature flags and use dark launching to safely clean up any other migration tasks required to leave the hosted model behind.

- Scheduled downtime is a relic of the past. There's no reason to take the site down to perform upgrades or make changes to the architecture. New services are launched and old ones are retired and the customer gets the same consistent user experience. High-performing SaaS offerings measure planned downtime in seconds.

- The software operates well in a single cloud across multiple data centers. Many companies worry about being in a single cloud because "What if Amazon goes down?" I generally advise that getting really good at running in one cloud is much more important than running in multiple clouds. If a company operates effectively in one cloud, it will be able to handle the occasional cloud provider problems, often within its cloud (but not necessarily in the same region). Companies that have successfully made the on-prem-to-SaaS transition should expect to operate successfully for two to four years before considering tackling the complexity of running multi-cloud, if the complexity of such operation makes business sense.

## *Business Outcomes*

Cost of goods sold is a major concern for portfolio companies that are trying to improve their profitability. There's a high cost of deploying software on-prem vs. a very highly leveraged SaaS model. While the hardware costs may or may not be borne by the customer, the level of complexity is enormous. Complexity is not free.

By reducing manual work, by having fewer emergencies, by having more satisfied customers, we can reduce the costs associated with these installations. We're able to take advantage of the rapid innovation available in the cloud to make further improvements in the architecture and operation of the service. This in turn brings in more customers, higher sales, and increased margins.

**DevOps Patterns for Private Equity**

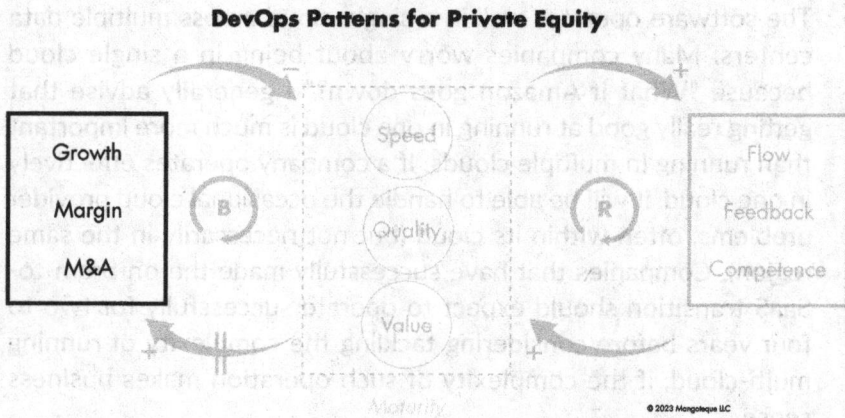

The migration from on-prem to SaaS requires almost all the elements of software delivery we've discussed throughout this book. Teams need to deliver on time, collaboratively, with both speed and quality, and be prepared to respond in the face of inevitable failures.

When this is achieved, there are fantastic outcomes that await, both professional and financial. The measures by which PE firms measure their portfolio companies are realized. Great exits are possible.

# What's Next?

In an interview published in early 2022,[50] top investor Orlando Bravo described the past 20 years of private equity in three phases:

- *2000–2010:* This period was when he really started to wonder why more people weren't investing in software companies. The numbers just looked so much better for investing in the software companies themselves rather than the audience these companies were serving. It was probably no coincidence that this was about the same time as the rise in SaaS, Salesforce being a prime example. He said this period was typified by the ability to buy cheap, cut costs, and increase prices.

- *2010–2014:* This was the period where leverage came in permanently. Because of that, much larger deals were possible, and it opened up a significantly larger part of the market to private equity.

- *2014–now:* He describes this period as all about growth. It's not about cutting costs or using leverage. In fact, he said that costs are only a small part of it. There are no cheap companies, or cost-cutting companies, left to buy.

Like the serendipity of the rise of SaaS in the early 2000s, there is serendipity today in the rise of DevOps. Private equity firms were uniquely suited to take advantage of the rise of SaaS, and in an era where growth is the name of the game, they're well suited to take advantage of the rise of DevOps.

With its holistic view of the software development process that aims to maximize human and technical capital, the name of the game for DevOps is to increase the chances of meeting business goals, including company growth. The company that can run more experiments can quickly deliver a higher quality service that customers want, and can do so without taking on massive risk or costs, is in a great position. As the *2019 State of DevOps Report* says: "Our highest performers are twice as likely to meet or exceed their organizational performance goals." Those

---

50  https://youtu.be/bbLgx4re37U

goals can include company growth, and what private equity executive or operator would not like to double their chances to meet or exceed their goals?

Similarly, in their *Harvard Business Review* paper "Private Equity's New Phase,"[51] Dave Ulrich and Justin Allen also describe three phases for private equity, though they don't agree with Orlando Bravo on the exact years. They argue that firms have moved on from "buy and sell" to "buy and transform."

*"But for phase three (buy and transform) ... operational excellence is not a tool, but a mindset. That means that in this phase, PE firms also require expertise into leadership, talent, and organizational capabilities and culture. Acquired organizations have organization capabilities that have to be transformed and talent that must be assessed and upgraded."*

We have been discussing these concepts throughout this book. Transforming organizational capabilities, upgrading talent (writing code!), and the fact that it's the leadership, not the engineers, who are responsible for making this transformation happen with a DevOps mindset. Leveraging the technology organization to achieve business outcomes.

They conclude that *"Research by accountants and economists ... has indicated that over 50% of the value of a firm cannot be explained by its financials and that new measures should be used to assess firm value."* Astute operators in private equity understand these lessons. I've performed many due diligence assessments for private equity firms that look at the software delivery capabilities of the targets. It's these firms that are positioning themselves to become or maintain their position as leaders in successfully delivering promised returns for their limited partners. It's these firms that begin trying to optimize the operation and delivery of the technology organization early in the holding period in order to maximize growth for that duration. No one wants to sell a

---

51  https://hbr.org/2016/08/private-equitys-new-phase

company that's struggling with the basics when those basics are now well understood.

It used to be that people would fly to California to talk about DevOps and lament that when they went back home, it was only a few forward-thinking Silicon Valley companies that were using it in practice. Today, practicing DevOps is the standard for how companies grow and operate. Many software companies purchased by private equity firms were founded before these DevOps principles had gained traction. They're at a disadvantage compared to their younger competitors.

By bringing DevOps into these organizations, by leveraging their considerable financial might and operational expertise, the days of being disadvantaged can soon pass. Happier employees, happier leadership, and happier investors. Win–win–win. *Viva la revolución.*

# Appendix A: Document Templates

## Sample Technical Specification

TITLE:
*Including month/year.*

SPECIFICATION AUTHOR:

STATUS: *(in progress/ready for comment/accepted)*

ABSTRACT:
*Summary/overview of what is being proposed. Basic explanation of scope and implementation. Familiarize those who are unfamiliar with what this proposal is all about.*

MOTIVATION:
*There is a problem we want to solve or a feature we need. Why?*

GOALS:
*Expected outcomes of the project, expected benefits of the project, who will it benefit?*

**PROPOSALS:**
Option 1
Approach #1
Option 2
Approach #2

**PHASES:**
*We always deliver functionality incrementally. What are the steps to get there?*
Phase 1
*This is the MVP.*
Phase 2
*More functionality/capability based on MVP learnings.*

**SUCCESS CRITERIA:**
*How do we know if we've achieved our goals?*

# Gameday Template

Date:

Test: *Sever connection to the database*
Expected behavior: *Graphs will show an increase in errors, JDBC connector will detect timeout after 30 seconds, alerting system will notify on-call responsible.*
Observed behavior:
Notes:

Test:
Expected behavior:
Observed behavior:
Notes:

Test:
Expected behavior:
Observed behavior:
Notes:

# Index

www.ingramcontent.com/pod-product-compliance
Lightning Source LLC
Chambersburg PA
CBHW011159220326
41597CB00026BA/4675